T0251334

INFORMATION THEORY AND ARTIFICIAL INTELLIGENCE TO MANAGE UNCERTAINTY IN HYDRODYNAMIC AND HYDROLOGICAL MODELS

To Atitegeb Alamir

Information Theory and Artificial Intelligence to Manage Uncertainty in Hydrodynamic and Hydrological Models

DISSERTATION
Submitted in fulfilment of the requirements of
the Board for the Doctorate of Delft University of Technology
and the Academic Board of the UNESCO-IHE Institute for Water Education
for the Degree of DOCTOR
to be defended in public
on Monday, 24 May 2004 at 15:30 hours
in Delft, The Netherlands

by

Abebe Andualem JEMBERIE

born in Merawi, Ethiopia

Bachelor of Science in Hydraulic Engineering (Arbaminch Water Technology Institute, Ethiopia)

Master of Science in Hydroinformatics with Distinction (IHE Delft, The Netherlands)

CRC Press
Taylor & Francis Group
Boca Raton London New York

CRC Press is an imprint of the
Taylor & Francis Group, an **informa** business

A BALKEMA BOOK

Published by:
CRC Press/Balkema
Schipholweg 107C, 2316 XC Leiden, The Netherlands

© 2004 by Taylor & Francis Group, LLC
CRC Press/Balkema is an imprint of the Taylor & Francis Group, an informa business

No claim to original U.S. Government works

ISBN 13: 978-90-5809-695-1 (pbk)

Visit the Taylor & Francis Web site at
http://www.taylorandfrancis.com

and the CRC Press Web site at
http://www.crcpress.com

Abstract

Recent decades have shown an increase in the application of mathematical models of water-based systems. Most of these models are based on the mathematical formulation of physical laws such as the conservation of mass and momentum. Such physically based models are widely applied in prediction and forecasting. However, there is also the growing importance of another class of models, called data-driven models, such as artificial neural networks, genetic programming, fuzzy logic generators, support vector machines, etc. in which the model is generated based on historical data alone rather than the physics of the particular application. With the increase in the application of physically based models, the associated uncertainty is becoming particularly important because, among other reasons, what the models predict and what is observed from the physical systems fit imperfectly. No mathematical model can be a perfect representation of the physical system it is intended to represent. This research intends to minimize the mismatch between physically based models and observations by the use of intelligent data-driven models and information theory-based principles.

This research is motivated by the fact that in developing physically based and data-driven models the modeller needs different resources to be made available, particularly physical insight and data, which indicates that they have a complementary nature and can therefore be optimally applied in some way. This leads to two basic assumptions. The first assumption is that the overall predictions of a physically based model can be greatly improved through the conjunctive use of a data-driven model of the residuals, and by so doing the associated uncertainty can be systematically reduced. The second assumption refers to the degree to which improvements can be made to the physically based model. The complementary data-driven model resolves a proportion of the residual error between the physically based model and observations. It does this by relating that proportion of the residual error to selected variables. The proportion of the residual error explained by the data-driven model is therefore an indirect measure of the potential that exists to improve the physically based model.

The second part of the thesis, chapters 3, 4 and 5, constitute the methodology developed and applied in the thesis. At first, information theory-based principles are investigated in relation to uncertainty. The average mutual information measure is found to be a very important tool in the research since it can be used to evaluate how much is known about one data set if another data set is already known. The average mutual information measure is further developed into a weighted mutual information measure, which enables the measurement of the mutual information in particular ranges of the data. Its application is demonstrated with the problem computing flood wave speeds using discharge hydrographs in rivers. In chapter 4, the basic principles of artificial neural networks and fuzzy set theory are investigated. These techniques are used to develop intelligent data-driven models in most of the case studies of the thesis. In relation to the principle of uncertainty invariance, it is demonstrated that fuzzy set theory and Monte Carlo simulation give similar information regarding the uncertainty in a ground water contaminant transport model that has uncertain parameters. Chapter 5 introduces the concept of complementary modelling that bridges the gap between a physical system and its (physically based) model by a complementary data-driven model. A complementary model can be used to predict the errors in forecasts made by a physically based model and subsequently to generate updated forecasts as well as forecasting the expected accuracy in the form of confidence bounds and linguistic measures that can be used as a measure of uncertainty. The methodology is tested with a hydrodynamic model of a hypothetical estuary, a river flood routing model and a conceptual rainfall-runoff model.

The practical applications of complementary modelling to existing operational models and to models that are developed under this study is demonstrated in the third part of the thesis. The first application is to flow forecast models for the Rhine and Meuse Rivers in The Netherlands. For both rivers, complementary modelling helped to improve discharge forecasts compared to forecasts made by neural network models and physically based models separately. It also helped to incorporate processes that are missing in the original model such as tidal effects for the Rhine model and lateral inflow for the Meuse model.

The second application deals with the Dutch Continental Shelf Model (DCSM), which is a 2-D model used to forecast water levels, surges and currents in the North Sea. The study done with the DCSM was different from the other case studies in that direct forecasting of errors with a complementary model failed to improve surge predictions significantly. Therefore the problem was reformulated such that the complementary model was trained to forecast the accuracy of physically based model and not the errors directly. The accuracy was defined using the bias and 90% confidence intervals as well as the linguistic descriptions of the expected errors. This demonstrated that a complementary model need not only provide a forecasting of the errors, but also can be trained to forecast the accuracy defined in some other form. In the same case study, the application of fuzzy rule-based models to characterize the accuracy of surge predictions in the form of IF-THEN rules is demonstrated. The rules relate known sea state or meteorological variables to the expected accuracy of the surge prediction in a linguistic form.

In addition to the possibility of forecasting the errors and accuracy of model predictions, the studies have indicated that, along with appropriate analysis techniques, patterns in model errors can be used as indicators to make further improvements to physically based computational models. For example, the errors in storm surge predictions of the DCSM consist of a dominant periodic pattern and its periodicity coincides with the tidal cycle. This indicates that it is necessary to sort out the non-linear relationship between the surge and the tide in a better way than is done at present. In an ideal situation, there should not be any periodicity in the surge prediction errors because the surge is not a periodic process. Recommendations for the improvement of models are given in other case studies as well.

The research has demonstrated that, besides its applicability without affecting the regular operation of the primary (physically based) model, complementary modelling offers a series of other advantages such as the possibility of incorporating data that are not used by the primary model and almost no demand for additional computer time. The possibility of obtaining information that can help in the improvement of the physically based model is also demonstrated, especially for the DCSM. The complementary modelling methodology is developed in a way that is not specific to a model of any particular situation. It can be applied to radically different models of physical systems besides those explored in this thesis.

Acknowledgement

My greatest thanks goes to my promoter Prof. Roland K. Price whose uninterrupted advice, support and ideas led to the success of this work.

I sincerely appreciate Dr. Dimitri Solomatine for the interesting conversations we had on many occasions. I am grateful to Prof. Mike J. Hall and Dr. Barbara Minsker for providing helpful comments and suggestions that helped in the improvement of the draft thesis.

I would like to express my sincere appreciation to UNESCO-IHE and its staff who are always there every time help is needed.

Parts of this work were financed by the NAUTILUS project of RIKZ. I am grateful to the financial support of RIKZ and the teamwork I enjoyed with Ir. D. Dillingh and Dr. M Verlaan.

Last but not least I take this opportunity to express my gratitude to my parents for their unconditional support in every step I took throughout my studies.

Table of contents

PART I. OVERVIEW

CHAPTER 1. INTRODUCTION

1.1. Problem description

In past decades, the engineering and scientific research community has been developing mathematical models of various physical systems in the pursuit of a proper understanding of the processes involved and the subsequent application of the models in making decisions. Water-based systems have been a particular area in which many advances have taken place. Subsequently significant progress has been made in the development and application of mathematical models, which have been in turn heavily dependent on advances in digital computing. The general trend in modelling has been from the development of tailored models of particular physical systems to generic modelling systems that can be used to instantiate models for a class of physical systems (Abbott, 1991). The result is that a large number of off-the-shelf modelling systems have been developed for both commercial and research purposes. Apparently, models developed using different modelling systems have different structure, data needs and, subsequently, accuracy in their representation of the physical processes under consideration. It is now the concern of most model users to make the right choice of modelling system for any particular problem. In this respect, the modeller should know what processes are included and how they are represented since, as it is stated in Cunge *et al.* (1980), the quality of the model results is a direct function of the modeller's effort to understand the problem. Part of the reason for this is that no mathematical model can ever be a perfect representation of the world it is intended to reproduce. All such models involve some 'uncertainty' because how the physical system actually behaves and how the models represent it always have some shortcomings about the goodness of fit.

For the purpose of this work, uncertainty associated with models refers to the imperfect and inexact representation of the 'real' world. Uncertainty occurs in data as much it does in the model or due to the inference mechanism used to represent the world under consideration. Data uncertainty has to do with what is measured or observed from nature. Model uncertainty has to do with how these observations are used to reason to a prediction, which prevails even while reasoning is done based on perfect data, that is, if perfect data ever exist.

This 'gap' between a physical system and its model is of particular interest to this thesis, which raises two important points of interest: (1) the presence and extent of any useful information on this gap, and (2) the possibility of bridging this gap and systematically reducing the uncertainty associated with it. This study addresses these two issues using information theory-based principles and intelligent data-driven modelling techniques. Even though the methodology is investigated with case studies focusing on physically based models of water-based systems, it is developed in such a way that it can be applied to mathematical models of any other physical system as well.

1.2. Deterministic models

1.2.1. Physically based models

Deterministic physically based computational models of an identified natural system are based on an understanding of the associated physical processes and their encapsulation in some form of symbolic equations. For almost all such natural systems the equations are derived under a number of simplifying assumptions, the primary aim being to ensure that the most important processes are included accurately, and that at least the 'gross' features of the corresponding phenomena are reproduced. The inclusion of every process within a tractable model is often not possible or even desirable. This may be due to an insufficient understanding of the processes involved or of appropriate ways to represent them. There may also be insufficient data from the real world to identify properly a process and its symbolic representation for a particular situation. When it comes to the computational aspects of solving the governing equations within the context of a specific physical boundary domain then a range of other problems arise. These include, for example, the reliability and robustness of the numerical solution of the governing equations, and the corresponding digital resolution of the physical domain.

A consequence of these assumptions and simplifications is that the resulting model based on the equations and their (numerical) solution within a specific context will always be an approximation of the real-world system. Of itself the model, which essentially is a reflection of how the modeller views the world, defines its own 'world' with its own specific (and limited) set of rules. The reason the model was conceived and developed, however, is to represent the 'real' world. Therefore the question arises as to how well, or how adequate is the match between the two worlds. Inevitably there are differences between the model predictions and the corresponding observations in the real world system. These differences, or residuals, can be termed 'errors'. The aim of the modeller is to reduce these residuals consistently such that the physically based computational model can be used with confidence for its intended purpose.

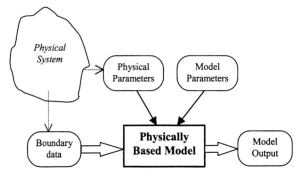

Figure 1.1. General setting of a physically based model

As it is depicted in Figure 1.1, physically based computational models generally have physical parameters to describe the physical domain and model parameters pertinent to the numerical solution. It is a general practice to 'calibrate' the instantiated model by refining the choice of values for certain parameters as a means of reducing the residuals. Care has to be taken to ensure that the values used are consistent with the theory developed for the particular processes being represented and therefore that the values lie within particular ranges. The very real danger is that forcing the values outside their acceptable ranges in order to minimize the residuals leads to a physically based model in which the symbolic representations of particular processes are pushed outside the valid limits of their derivations. Application of such a model to scenarios where the physical boundary domain is altered can then generate results that are not safe or reliable to use. In this sense it is better to limit the refinement of the confirmed model and to work with a proper understanding and appreciation of the uncertainties generated by the residuals. Moreover, the calibration of model parameters may not account for all systematic patterns that prevail in the residual errors. The problem becomes even worse when there are defects in the model structure. Some interesting points concerning the issue of model calibration are presented in Cunge (2003) with examples of river and estuarine models.

1.2.2. Conceptual models

Another class of models that are in common use are conceptual models. Conceptual models are generally based on some conceptual analogy of the physical processes under consideration. They also maintain some basic physical laws such as mass conservation. In this regard, these models require physical insight. Their other characteristic feature is that they are generally over-parameterised. Most if not all of their parameters are not measurable and are determined by calibration using relevant time series data. Catchment runoff modelling is a typical example where conceptual models are widely applied. An example of a conceptual analogy in catchment runoff modelling is the assumption of a catchment as a series of linear reservoirs, which is often referred as the Nash-cascade (Nash, 1959).

1.2.3. Data-driven models

In recent years, there has been the growing importance of another class of models, called data-driven models such as artificial neural networks, genetic programming and fuzzy logic generators (see for example Kasabov, 1996). Unlike a physically based computational model, a data-driven model is not based on an explicit representation of discrete physical processes but on the adoption of a particular technique trained to connect selected causative and resultant (input and output) sets of data. Physical insight is necessary to select the appropriate data sets and even to oversee the process of constructing and training the connection between the two sets. However, such modelling techniques are generally independent of the physics of the particular application.

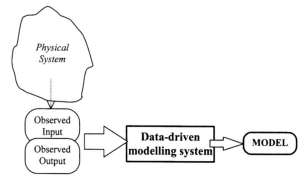

Figure 1.2. Data-driven modelling

In general, provided there are sufficient data, a data-driven model is easier to set up for routine operation. It can also be better in reproducing the observed resultant data than a physically based computational model. Despite their simplicity, data-driven models are limited to learning from the data they are provided with. So if there are intended changes to the physical boundary domain, the data-driven model is unable to reflect these changes unless there are corresponding causative and resultant data sets available. Similarly data-driven models are less able to predict accurately the resultant data when the causative data is outside the range for which the model is trained. Like the physically based model, the data-driven model is also an approximation of the real world situation.

Recent practice shows that there are several shareware and commercial software tools available to develop data-driven models. Some of these data-driven modelling systems offer a range of intelligent learning or *training* algorithms to generate models from historical data.

As depicted in Figure 1.2, the model is essentially an *output* of the learning operations of such a modelling system. Examples of such systems are NeuroSolutions and WEKA.

1.3. Basic philosophy and hypothesis

In retrospect, the description given in §1.2 regarding different classes of deterministic models shows that there are two resources to develop such a model: the physical insight about the involved processes and the historical data collected from the physical system. It also shows that these classes of models have a basic difference with respect to how much of each resource is needed to develop them. Abebe & Price (2004) used the diagram shown in Figure 1.3 to illustrate the relative position of these classes of models in the spectra of the demand for physical insight and historical data. Accordingly, physically based models are dependent on physical insight whereas data-driven models depend on historical data as their names imply; they thus lie at opposite ends of the spectrum. Conceptual models lie in between the two, their relative position depending on factors such as the extent of physical knowledge they incorporate, the number of parameters they have and, consequently, the minimum acceptable amount of historical data needed to determine the values of their parameters. For the purpose of this thesis, conceptual models are treated as being physically based.

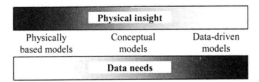

Figure 1.3. Classes of models as viewed in the spectra of physical insight and data needs (adapted from Abebe & Price, 2004)

Another deduction from the diagram in Figure 1.3 is that physically based models and data-driven models have a potential *complementary nature* in that they are derived from different directions. This very difference and the resulting complementary nature of these two classes of models is also reflected by the complementary nature of their advantages and disadvantages which leads to the notion that they can be effectively used together in an optimal manner. This complementary nature is the philosophical basis used to develop of the *complementary modelling* approach presented in this thesis, which is based on using a data-driven model as a complement to a physically based computational model.

Two hypotheses are used to formulate the methodology. The first hypothesis is directly based on the complementary nature of physically based computational and data-driven models and can be stated as:

❑ The overall predictions of a physically based computational model can be greatly improved through the conjunctive use of a data-driven model of the residuals, and by so doing the associated uncertainty can be systematically reduced.

The second hypothesis is a sort of a 'feedback' mechanism that reflects on the physically based computational model itself and is essentially the inverse of the first hypothesis. It can be stated as:

❑ The degree to which improvements can be made to the predictions generated by a physically based computational model through the application of a complementary data-driven model is an indirect measure of the potential room available to make improvements to the physically based model.

1.4.   Objectives of the thesis

The objectives of this thesis are:

❑ To investigate the possibilities of applying information theory to evaluate the amount of information contained in historical prediction errors of physically based computational models by relating them to variables that reflect the state at which the predictions are made such as input data, state variables, model output, and other data that are relevant but not used by the model.

❑ To develop a methodology to manage the prediction uncertainty of a physically based computational model by the use of intelligent data-driven models to bridge the gap between the physically based computational model and the corresponding real world prototype on the basis of the complementary model concept. The methodology is applied to different physically based computational models by connecting particular variables to the prediction accuracy.

❑ To attempt to use patterns in the historical model errors as indicators that suggest possible ways of improving the physically based computational model by combining physical insight with appropriate data analysis techniques.

1.5.   Outline of the thesis

The thesis is broadly divided into four parts: overview (chapters 1 and 2), methodology (chapters 3, 4 and 5), application (chapters 6 and 7) and evaluation (chapter 8).

Chapter 2 provides an overview of different sources of uncertainty pertinent to physically based computational modelling. It reviews the various theories used to conceptualise uncertainty in different forms. It briefly summarizes some of the most commonly used techniques of analysing uncertainty in mathematical models.

Chapter 3 discusses the basic principles of information theory. It particularly focuses on the relationship between uncertainty and information. A new form of a biased information measurement approach is developed, and its application is demonstrated through a problem involving the relationship between flood wave speed and discharge in rivers.

In chapter 4, artificial intelligent techniques, which are used in developing data-driven models for case studies presented in the rest of the thesis, are introduced. The first part covers artificial neural networks with a focus on multilayer perceptron and radial basis function networks. The second part covers techniques based on fuzzy set theory and presents a case study that compares fuzzy alpha-cut technique and Monte Carlo simulation in analysing parameter uncertainty in a ground water contaminant transport model.

Chapter 5 brings the pieces of the methodology together by introducing the concept of complementary modelling of physically based computational and data-driven models. Techniques based on information theory are used to trace back recoverable information in the historical errors generated by models by relating them to the model state variables. Intelligent data-driven models techniques are used develop a model for the 'residual process'. The methodology is further demonstrated with hypothetical and real life case studies in hydrodynamics, river flood routing and rainfall-runoff modelling.

In chapter 6, the complementary modelling approach is applied to a problem of forecasting flows on the Rhine and Meuse rivers in The Netherlands. Both data-driven and physically based computational models are developed. A comparison is made between neural network, physically based and complementary modelling techniques. It demonstrates that a proper analysis of residual errors give valuable information that indicates particular misrepresented processes that can be addressed by complementary modelling.

In chapter 7, the methodology is extensively applied to coastal surge forecasting with a 2-D hydrodynamic model. Information theory is used to analyse the time dynamics and information flow of the surge and its prediction errors in relation to selected meteorological and sea state variables. Subsequently, complementary modelling is applied to establish neural network models of bias and confidence intervals of the predicted surge with a remarkable accuracy. It also addresses the problem of relating state variables to the expected surge prediction accuracy using genetic algorithms along with a fuzzy rule-based technique, which is a new approach developed in this study. An attempt is also made to obtain valuable information that indicates possible ways of improving the original model.

Chapter 8 summarizes and concludes the important findings of the research based on a broader view of the various case studies presented in the thesis. Finally, suggestions of possible directions for future research are given.

CHAPTER 2. BACKGROUND

This chapter is intended to provide background information on various sources of uncertainty that exist in physically based computational models and the different ways it affects model performance. It provides a brief literature review on the different theories used to represent uncertainty in alternative forms. It classifies model uncertainty in a characteristic form – based on the way its effect is revealed in the model prediction errors. Such a classification is chosen since it relates to the methodology developed and applied in this thesis. A summary of the most commonly used techniques that are used to analyse uncertainty in prediction models is also presented. The purpose of the chapter is to provide the reader with an overview of the terms that are stated in the rest of the thesis.

2.1. Uncertainty in physically based models

Modelling a physical system involves uncertainties that arise from various sources. The implication of these uncertainties is particularly important when the results of the model are used in some sort of decision-making process. For a proper application of models, the associated uncertainty needs to be *analysed,* meaning that its presence and extent has to be detected, *represented,* meaning that it has to be described in a form that is understandable and, if possible, *reduced* so that the quality of information obtained from the inferences made by the model is improved.

A systematic uncertainty analysis provides insight into the level of confidence in the model results, and can assist in the assessment of how reliable model predictions are. Further it can lead to the identification of the key sources of uncertainty that are or are not important with respect to the purpose for which the model is developed. The purpose of uncertainty analysis is to use the available information in order to evaluate the level of confidence in the existing data and model. This alone will not reduce uncertainty, because reduction of uncertainty can only come from gathering additional information and filling gaps both in data and model.

For any such analysis, it is important to know just what the sources of uncertainty in physically based computational models are. Uncertainty can result from the data as well as the model. The uncertainty arising from the two sources are discussed below as *data uncertainty* and *model uncertainty* respectively.

2.1.1. Data uncertainty

Uncertainty related to observed data can be classified in two ways: measurement uncertainty and sampling uncertainty. *Measurement uncertainty* refers to the actual quality of the data with respect to the precision of the measuring equipment and the process of measurement, including the human factor. *Sampling uncertainty* refers to the temporal and spatial resolution of the data sampling because most mathematical models are defined to work based on discrete instead of continuous data, which raises the question whether the resolution of the measurement describes the quantity in sufficient detail.

Data observed from the physical system are used as boundary data, also called model forcings, and physical parameters, which are static data that define the model domain. These data are subject to human and instrumental errors even when the most accurate measurement techniques are applied. The observed equivalent of the model output is also subject to measurement errors.

Errors in the physical parameters that define model domain obviously remain as part of the model. The effect of uncertainty associated with the quality and spatial and temporal resolution of boundary domain data (or model forcings) can be seen at two stages of the modelling process: model development and model application. At the model development stage, model parameters that cannot be directly measured are generally calibrated based on the goodness of fit between observed data and corresponding model predictions on the calibration data. The use of non-representative data affects the calibration of the model parameters and the calibrated values remain as a part of the model. At the stage of model application, errors in observation translate through the model to its output. In a related context, Mroczkowski *et al.* (1997) stressed the importance of incorporating data with a response different for the one used in determination of parameters and using data other than stream flow in the validation of rainfall-runoff models to obtain more representative parameters and model structure.

2.1.2.   Model uncertainty

Mathematical models can be considered as simplified representations of rather complex physical processes in the physical system. An ideal mathematical model is one having great simplicity while representing the involved processes with adequate accuracy. Model uncertainty here refers to the uncertainty in the inference mechanism used by the model such as its structure, spatial or temporal resolution (e.g. numerical grid cell size), simplification, ignored or misrepresented processes, interaction with the outer world, uncertain model parameters, and extrapolation. There is no clear distinction between some of these sources of uncertainty, which are described below.

Structural uncertainty

The structure of the model employed to represent the physical system is often a key source of uncertainty. Uncertainty arises when there are alternative sets of scientific or technical assumptions for developing a model. In such cases, if the results from competing models result in similar conclusions, then one can be confident that the decision is robust. If, however, alternative model formulations lead to different conclusions, further model evaluation might be required. Conceptual rainfall-runoff models are a typical example of models with different structures. Franchini & Pacciani (1991) compared six conceptual rainfall–runoff models that use different structures to represent the involved processes. Their results showed the variation in performance when applied to the same data.

Simplification, ignored and misrepresented processes

In the process of modelling, the mathematical formulations involved are usually simplified. Simplification is often used as a means of overcoming the unavailability of data, complexity of the actual physical system, lack of knowledge of all the involved processes, or reduction of computational burden. In some cases, simply the gain of accuracy obtained by including some of the processes is not worth the resulting complexity of the model and those processes are ignored. Most mathematical models therefore focus on the major processes. This makes the resulting models to deviate further from the represented physical system. The effect of each process ignored or misrepresented while setting up a model contributes to the discrepancy between the model outputs and the observations.

Parametric uncertainty

A mathematical model consists of parameters that represent some aspects of the physical system or the numerical solution employed. The parameters may or may not have a direct physical meaning. They also may or may not be obtained by direct measurement. When direct measurement is not possible (or difficult) their values are usually determined by other means such as expert judgment, model calibration or comparing with values obtained experimentally; for example, from a scale model. Expert judgment lacks numerical precision. Calibration involves uncertainty in that it may force the values of the parameters to be unreal especially when the calibration data is not representative and model structure is not good. In addition, the computational load of some models could be prohibitive to perform calibration. The use of experimental values of parameters is also subject to uncertainty since exact replication of the physics is almost impossible. Therefore, the values of model parameters are subject to uncertainty.

Interaction with the outer world

The physical system represented by a model is generally a component part of a much larger physical system. A part of a complex physical system is separated for modelling for various reasons. This separation is done in such a way that any significant interaction with the 'outer world' is represented as a boundary condition to the model. A typical example is a hydrodynamic model of an estuary that incorporates the delta part of a river and part of a sea. In developing such a model, generally, the river boundary goes as far upstream as possible to exclude the effect of tides. The sea boundary goes as deep as possible to exclude the effect of the river currents and shallow bathymetry from the tidal boundary. Once the model domain is defined as such, the next step is to define how it interacts with the outer world. For the estuary model, the hydrodynamics could be affected by tidal waves, wind shear, surface pressure, and flow from the river, which are what most models consider. However, there are also other factors that can affect the hydrodynamic behaviour no matter how minor their contribution is, such as precipitation directly over the water body, temperature and density driven currents, interaction with groundwater, etc. Also in most cases, water quality computations (temperature, concentration, etc.) are done using the results of hydrodynamic simulations. However, the physical relationship between some of the processes affecting water quality and those affecting the hydrodynamics is a two-way interaction.

Obviously, taking all of these interactions as boundary conditions leads the modelling process to unimaginable complexity. Therefore, it is not hard to imagine that some interactions are considered as being insignificant and ignored; for some of these interactions, the necessary data is not available, while other interactions are yet unknown. The interactions that are not accounted for affect the model performance and can appear as part of the discrepancy between observed and model predicted values. These interactions are far from 'random'. In some cases, the interactions might just have different, much larger or smaller, time and length scales than what the modeller is interested in. The effect of these interactions is often termed as *dynamic noise* (see for example Droste, 1998; Daw *et al.*, 2003). This brings uncertainty into the model since the extent to which the interaction affects the model might not be known.

Schematisation, spatial and temporal resolution of the model

The numerical solutions of most models that are based on partial differential equations involve the selection of temporal and/or spatial grid sizes. This stage involves uncertainty, on one hand because finer grids, which are logically more representative, involve higher computational load, and on the other hand because there is a practical limitation on the validity of the governing equations of the model at finer scales. For most such models, there are established criteria to select appropriate spatial and temporal resolutions in order to bring the numerical solution as close to the analytical solution as possible. However, these criteria are generally defined on the basis of the state variables, which keep on changing with time. An example of such a change is the criteria based on the Courant number in free surface flow models, which, if it deviates from unity, subjects the numerical solution to amplification and phase errors (see Abbott, 1979; Cunge *et al.*, 1980).

Extrapolation

Due to constraints in the availability of representative data, models are validated for some portion of the boundary domain data. These models may turn out to be inappropriate for the prediction of events resulting from input data beyond the range used for validation. For example, there is uncertainty involved in applying a river model that is validated on the basis of observations limited to channel flow conditions to flows that involve floodplains even if the physical parameters of the model incorporate the floodplain as well.

2.2.   Conceptual bases of uncertainty

Various approaches are used as a conceptual framework to represent uncertainty. These approaches generally stem from some associated theory. Klir (1994) presents the different theories that are used as conceptual bases for representation of uncertainty. Some of these theories are summarized as follows:

Classical set theory uses mutually exclusive alternatives in situations where one alternative is desired. Here, the uncertainty arises from the non-specificity inherent in each set. Large sets result in less specific predictions than smaller sets. Full specificity is obtained only when one alternative is possible.

Probability theory expresses uncertainty in terms of a measure on subsets of a universal set of all possible alternatives. The uncertainty is described by assigning a number between 0 and 1 to each alternative outcome, which is known as the probability or likelihood that the desired alternative is in this subset. For a universal set U with a subset A the probability function has the property

$$p(A) = 1 - p(A')$$

where set A' is the complement of set A. Each subset of the universal set is assumed to consist of only one alternative. The subsets are also mutually exclusive, therefore each alternative assumes a non-zero probability. Probability theory is the basis of most uncertainty analysis techniques such as Monte Carlo simulation.

Fuzzy set theory is based on the use of fuzzy sets, which are a more general form of classical sets (Zadeh, 1965). In fuzzy set theory, subsets of the universal set are not mutually exclusive. This is because members can belong to more than one subset to a degree between 0 and 1, which is defined by the membership function. Fuzzy set theory is suitable to represent vagueness, for instance, uncertainty in linguistically described data such as "high" and "low". A detailed account of fuzzy set theory and associated techniques is given in Chapter 4 of this thesis.

Rough set theory is based on defining uncertainty by the use of finite ranges with a lower bound and an upper bound. Interval arithmetic is an example of an uncertainty analysis technique that is based on rough set theory. It assumes uncertain parameters as unknown but bounded. This assumption is particularly useful when the fuzzy membership function or the probability distribution of an uncertain parameter cannot be defined. A difference between fuzzy set theory and rough set theory is that in rough set theory, the 'membership' is assumed invariant (unity) within the fine range.

Evidence theory also known as Dempster-Shafer theory (Shafer, 1976) is based on two non-additive probability measures, which are *belief* and *plausibility*. Assuming a finite universal set U with a subset A, the belief function can be defined as:

$Bel: p(U) \rightarrow [0,1]$ such that

$Bel(\emptyset) = 0,\ Bel(U) = 1$ and

$Bel(A) + Bel(A') \leq 1$

which shows that the belief function is sub-additive.

The plausibility function can be defined as:

$Pl: p(U) \rightarrow [0,1]$ such that

$Pl(\emptyset) = 0, Pl(U) = 1$ and

$Pl(A) + Pl(A') \geq 1$

which is a super-additive function.

The two functions can be related as

$Pl(A) = 1 - Bel(A')$

Assuming that there is insufficient evidence regarding A, $Bel(A)$ expresses all our reasons to believe in A whereas $Pl(A)$ expresses how much we should believe in A provided that all the available facts are in support of A. The main advantage of evidence theory is that it gives room for doubt in the form of the difference between the plausibility and belief measures. Hall and Davis (1998) and Hall (1999) applied non-additive probabilities in a problem of managing uncertainty in coastal defence systems.

2.3. Characteristic classification of uncertainty

In relation to the various theories used to conceptualise uncertainty, Klir & Wierman (1998) classify uncertainty in three types: *fuzziness* (or vagueness) which results from imprecise boundary sets; *non-specificity* (or imprecision), which is connected with sizes (cardinalities) of relevant sets of alternatives; and *strife* (or discord), which expresses conflicts among various sets of alternatives. They further classify the latter two types as *ambiguity*. They also gave a detailed description of how these types of uncertainty are measured in the framework of different theories. In the literature, uncertainty has been defined and classified in different forms. Hall (1999) did a comprehensive review on the classification of uncertainty. In general, any classification is dependent on the problem in question and how the individual views it. In this respect it gives sense to the comment of Laviolette & Seaman (1994) that 'uncertainty is in the eye of the beholder'.

The classification of uncertainty in this thesis diverges from the above three classes. As stated in Chapter 1, the subject of this thesis is the gap between physical systems and their corresponding (physically based computational) models. At least in present practice, the best quantitative measure of the gap between a physical system and its computational model is the residual error between the prediction made by the model and the corresponding observed value. All the sources of uncertainty in physically based computational models discussed in sections §2.1 contribute to this gap. This gap is also the 'uncertainty' that it is intended to reduce. This does not mean that errors and uncertainty are the same. Also residual errors cannot be used as a measure of the whole gap between the model and the physical system since the availability of such errors is limited to scenarios under which the model is used and the errors that are generated. Nevertheless, model errors are not just numbers. They contain valuable information regarding the combined effect of all sources of uncertainty in model and data. However, to unlock and interpret this information, proper analysis techniques have to be used and the circumstances under which the model is used have to be known. In this way, errors can be used as an indication of the overall uncertainty involved as a result of the combined effect of all sources of uncertainty.

From this perspective, another criterion is used to classify or rather *sub-divide* this uncertainty. This criterion is based on the characteristics of the gap between the model and the physical system as it is revealed in the form of errors, which is essentially a *characteristic classification* that divides uncertainty into *structured* and *unstructured* forms.

2.3.1.   Structured uncertainty

Some portion of the uncertainty arising from factors such as model structure, parameters, systematic errors in boundary domain data and dynamic noise from the interaction between the model domain and the real world that is not accounted as a boundary condition has a pattern that can be accounted for deterministically and reduced systematically. It is called here 'structured' uncertainty because it follows a structure of its own which, provided that proper analysis techniques are applied, can be systematically revealed, represented by a (deterministic) model of its own, and subsequently reduced. Intuitively, there exists an ideal model to represent any structured uncertainty on the provision that all the necessary causative data are available.

2.3.2.   Unstructured uncertainty

'Unstructured' uncertainty is characterized by the absence of any pattern whatsoever in its occurrence. It can be caused by noise in measurement errors and human errors, for example. Also model errors induced as a result of the interaction of the model domain with the outer world can be a source of unstructured uncertainty if this interaction has a much smaller time scale than the main processes in the model. Its main feature at a conceptual level is that it is not possible to model unstructured uncertainty in a deterministic form. Its extent can however be indicated using confidence intervals or other stochastic schemes.

Obviously, the causes of the above two characteristic classes of uncertainty are not mutually exclusive. However, the contribution of a source of uncertainty to the gap between the real world observations and the corresponding predictions of the physically based computational model can be anticipated using some clues. For instance, if the boundary data of a model is considered, a low signal to noise ratio in the data is suggestive of unstructured uncertainty whereas a systematic error as a result of miss-calibrated measuring equipment can predispose a model to have a structured uncertainty. Structured uncertainty and unstructured uncertainty are in a way analogous to bias and noise in errors.

2.4.   Some uncertainty analysis techniques

Sensitivity analysis

Sensitivity analysis is essentially a way of testing how the output of a model responds to variations in its parameters or input data. It can be posed at different levels of complexity, such as varying a group of parameters all at once, often depending on what is sought from the analysis. It gives valuable information such as which data is worth measuring precisely and which data need more sampling points.

Monte Carlo simulation

Monte Carlo simulation is one of the most widely used techniques to analyse uncertainty due to input data and model parameters. In Monte Carlo simulation, an input parameter P subject to uncertainty is considered as a random variable **P**. A number of realizations P_i of **P** is generated and the deterministic model is run for each of them, hence producing an output R_i.

The set of outputs R_i represents the set of realizations of the random variable **R**. The statistical properties of **R** are therefore computed from the realizations R_i (see the illustration in Figure 2.1).

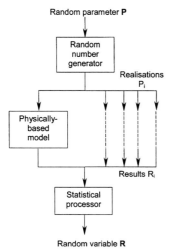

Figure 2.1. Monte Carlo simulation

An application of the Monte Carlo simulation technique is given in §4.4. In the case when there is a correlation between two or more uncertain parameters, the correlation can be incorporated during the generation of samples. The technique is widely applied in many disciplines. It generally requires a large number of samples (and model runs), which sometimes limits its applicability to simple models.

Latin Hypercube sampling

Latin Hypercube sampling (see, for example, Stein, 1987) is a variant of the standard Monte Carlo simulation technique. In this method, the range of probable values for each uncertain parameter is divided into segments of equal probability. Thus, the whole parameter space, consisting of all the uncertain parameters, is partitioned into cells having equal probability (but not necessarily of equal size). It enables an efficient sampling in that each parameter is sampled once from each of the segments. Its other advantage is that the random samples are generated from all the ranges of possible values; thus it enables extreme values of uncertain parameters to be represented, which could otherwise be ignored in the standard Monte Carlo sampling.

Generalized Likelihood Uncertainty Estimation

The Generalized Likelihood Uncertainty Estimation (GLUE) approach is an approach to incorporating uncertainty in real-time forecast models (see Beven & Binley, 1992). The GLUE approach rejects the search for a best set of parameters that gives a maximum goodness of fit. Rather, the approach evaluates the ranges of feasible parameters after perturbing the data for random errors in observations, and then uses a Bayesian framework to update the distribution of predictions as new observations become available. Romanowicz & Beven (1998) applied GLUE to a problem of real-time prediction of flood inundation probabilities on the River Culm, UK. It was reported that the approach helped to narrow the *a priori* predicted inundation probabilities as new observations come in.

Fuzzy alpha-cut technique

The fuzzy-alpha cut technique uses fuzzy set theory to represent uncertain model parameters. The model is then evaluated using the fuzzy parameter to determine the corresponding fuzziness of the model output. Since fuzzy membership functions cannot be directly used in most models, a form of interval arithmetic is used along with multiple model runs to construct the membership functions corresponding to the model outputs. The fuzzy alpha-cut technique is described in much more detail in §4.2.5 along with a case study on ground water contaminant transport problem (see also Abebe *et al.*, 2000b).

PART II. METHODOLOGY

CHAPTER 3. INFORMATION THEORY-BASED APPROACHES

3.1. Introduction

This chapter gives an overview of information theory-based principles that are linked to uncertainty. The need for exploring information theory is justified by the fact that uncertainty is related to information, or rather, the lack of it. If all the information about a particular physical system were available, there would be no uncertainty regarding that system.

Most applications of information theory in water-based systems are related to time series forecasting of chaotic systems that can be characterized by their high sensitivity to initial conditions and decreasing predictability with increasing forecast horizons. Applications to estuarine and coastal waters have demonstrated by Frison *et al.* (1999); Frison (2000); Velickov (2002).

Here an attempt is made to explore the link between information and uncertainty. Information theory-based measures are used to determine the presence and amount of information that can be used to manage the prediction uncertainty of models. The techniques explored here are largely exploited in the case studies and application problems presented in the later chapters of the thesis. Existing approaches are modified to suit particular application problems. The application of a modified information theory-based measure in a problem involving the relationship between flood wave speed and discharge in rivers is also demonstrated.

3.2. Uncertainty and information

Uncertainty and information are inherently tied concepts. Klir & Wierman (1998) stated that different types of information deficiencies such as imprecision, non-reliability, vagueness, contradiction or deficiency may result in different types of uncertainty. They also argue that, if the amount of uncertainty can be measured by some mathematical theory, the amount of information obtained by some action (finding a new fact, new relevant observation, or discovering a relevant historical record) may be measured by the reduction of uncertainty that results from the action. Klir & Wierman (1998) refer to information conceived in terms of uncertainty reduction as *uncertainty-based information*.

Information that can help in the reduction of uncertainty is of particular interest in this thesis. Measuring information from the point of view of reducing uncertainty has two obvious characteristics: (1) the measure is incomplete since it does not measure all the information available from the source unless it contributes to the reduction of uncertainty, and (2) it is an indirect measure since it measures the amount of information in terms of the amount of uncertainty it can resolve.

Direct measures of the amount of information without a link to uncertainty are also investigated in the literature. For example, from a point of view of computability, the amount of information represented by an object is measured by the length of the shortest program written in a standard language (e.g. for the standard Turing machine). This type of information is referred to as *descriptive or algorithmic information* (Kolmogorov, 1965; Chaitin, 1987).

There are various theories to relate uncertainty and information, namely, probability theory, possibility theory, evidence theory, and fuzzy set theory. Information theory as defined in terms of probability theory by Shannon (1948) will be used as a framework to relate information and uncertainty. The following simple experiment demonstrates the relationship and hierarchy between uncertainty and information in a probabilistic framework.

Three boxes (a), (b) and (c) containing red (R) and blue (B) balls are considered.

   (a) 1 R ball and 1 B ball

   (b) 1 R ball and 9 B balls

   (c) 1000 R balls and 999000 B balls

(E1) If a ball is taken out of each box, the uncertainty of knowing the colour of the ball is high in (a), less in (b) and least in (c).

(E2) Introducing some changes, if (c') contains 1001 R balls and 998999 B balls and (c") contains 1002 R balls and 998998 B balls, then, intuitively, (c') and (c") have the same uncertainty as (c).

(E3) If in box (b') there are 1 B and 9 R balls, the uncertainty is still the same as (b).

Next, the information obtained from removal of one ball from a box is considered.

In case (a) removal of a ball from the box gives more information no matter what the colour of the ball is. In cases (b) and (c) if the ball is B, less information is obtained since that is what was expected. On the other hand, if the ball is R, it gives more information since it was not expected. Therefore, if the outcomes of an experiment are evaluated in a probabilistic framework, uncertainty is an *a priori* concept, while information is *a posteriori* one.

3.3.   Entropy and uncertainty

The word 'entropy' comes from the Greek word for 'transformation'. Entropy is considered to be a quantity associated with the state of a system. Therefore it can be treated as any physical quantity despite the fact that it does not obey the laws of conservation. Its distinct nature though is the fact that it cannot be measured; rather it is determined statistically. Entropy is used as a measure of the degree of disorder or uncertainty within a system. Indirectly, it also reflects the information content of space-time measurements.

Formalization of the concept of Entropy

In order to formalize the concept of entropy, Shannon (1948) considered an experiment with n possible outcomes $x_1, ..., x_n$. Each result x_i has a probability of occurrence p_i (with $\Sigma p_i = 1$, $1 \leq i \leq n$). The target is to find a function $H(p_1, ..., p_n)$ which measures the uncertainty in performing the experiment.

Shannon established that this function must satisfy the following three conditions:

(C1) $H(p_1, ..., p_n)$ should be continuous in $p_1, ..., p_n$.

(C2) If all the p_i are equal, then $p_i = 1/n$. It follows that H should be a monotonically increasing function of n. With equally likely events there is more choice, or uncertainty, when there are more possible events.

(C3) If a choice is broken down into two successive choices, the original H should be the weighted sum of the individual values of H.

Subsequently Shannon gives a theorem which states that the only H satisfying the three assumptions above is of the form shown in equation (3.1),

$$H = -k \sum_{i=1,n} p_i \log p_i \tag{3.1}$$

in which H is a measure of information, choice and uncertainty.

Besides these three conditions, H also satisfies the following condition:

(C4) The value of $H(p_1, \ldots, p_n)$ does not change when the arguments $p_1, \ldots, p_n$ are permutated.

Referring to the experiment in the preceding section, condition (C1) corresponds to (E2) in that subtle changes in balance do not cause much change in uncertainty. Condition (C4) corresponds to (E3) in that interchanging the number of Red and Blue balls does not change the uncertainty.

It has also been proved elsewhere that there exists a function which satisfies these conditions, and this function is *unique* (see, for example, Klir & Folger, 1988). The function is called Shannon entropy.

When entropy is applied to a random variable (X,Y) whose values are in the set $\{(x_i, y_j), x_i \in A; y_j \in B \}$, it has the following properties:

(1) The entropy of the random variable (X,Y) is equal to the sum the entropies of X and Y/X (Y given X):

$$H(X,Y) = H(X) + H(Y / X) = H(Y) + H(X / Y) \tag{3.2}$$

where H(Y/X) is the conditional entropy of Y given X.

The entropy of the random variable (X,Y) is given by:

$$H(X,Y) = -k \sum_i \sum_j p_{ij} \log p_{ij} \tag{3.3}$$

where $p_{ij} =$ is the probability of $\{X = x_i, Y = y_j\}$

(2) The entropy of the random variable (X,Y) is always less than or equal to the sum of the entropies of each of the random variables X and Y:

H(X,Y) ≤ H(X) + H(Y)

(3) H(X,Y) = H(X) + H(Y) if and only if the random variable X and Y are independent.

As it can be noticed in equation (3.1), the entropy is an average measure, more specifically, a weighted average in which the weights are the probabilities of each individual outcome. The amount of information from a single event causing $\{X=x_i\}$ is defined by

$$I(x_i) = -\log p_i \tag{3.4}$$

Kantz & Schreiber (1997) maintain that the inverse of the numerical value of the entropy of a time series is the time scale relevant for the predictability of the system. The Shannon entropy is well defined for data of discrete nature such as symbol strings; however, it is not a unique characteristic for an underlying time series. The reason for this is that its value depends on the choice of symbolic encoding, for instance the value of the thresholds chosen. This implies that it is not invariant under smooth coordinate changes, not even under a simple change of experimental units.

3.4. Average mutual information

The *average mutual information* (AMI) or mutual information, which is based on Shannon's entropy, is the measure of information that can be learned from one set of data having knowledge of another set of data. According to Shannon, the information that a random variable Y gives about the random variable X is equal to the amount by which Y reduces the uncertainty about X, namely:

$$I(X;Y) = H(X) - H(X/Y) \tag{3.5}$$

It is a 'mutual' information and not just information, because rearranging equation (3.2) gives:

$$H(X) - H(X/Y) = H(Y) - H(Y/X) \tag{3.6}$$

which implies

$$I(X;Y) = I(Y;X) \tag{3.7}$$

Therefore, X gives the same amount of information about Y as Y gives about X.

Comparing equations (3.2) and (3.7), the AMI can be written as:

$$I(X;Y) = H(X) + H(Y) - H(X,Y) \tag{3.8}$$

The importance of the AMI became clear after Fraser & Sweeney (1986) used it to determine the optimal time delay in time series prediction as a better alternative to the autocorrelation function. The autocorrelation function, which is rather well established in time series analysis, has practical advantages in that it helps to become acquainted with the data and gives ideas about stationarity and typical time scales. The main objection to it is that it is based on linear statistics (see equation (3.12)) and hence does not take into account non-linear dynamical correlations. The AMI does not depend on any particular function and therefore can help to detect both linear and non-linear correlations.

To compute the AMI between a time series $S(t)$ and its time delayed copy $S(t+\tau)$, a histogram is created to determine the probabilities on the interval that the data covers. If p_i is the probability that the signal assumes a value inside the bin i of the histogram, and let $p_{ij}(\tau)$ is the probability that $S(t)$ is in bin i and $S(t+\tau)$ is in bin j, then the mutual information for time delay τ is:

$$I(\tau) = \sum_{i,j} p_{ij} \log p_{ij}(\tau) - 2\sum_i p_i \log p_i \tag{3.9}$$

Equation (3.9) is obtained by substituting the equations for entropy (3.1) and (3.3) in equation (3.8). Since equation (3.9) refers to the same time series, H(X) and H(Y) will assume the same value.

Theoretically, in the limiting case of large τ, $S(t)$ and $S(t+\tau)$ are independent and $I(\tau)$ becomes zero. Fraser & Sweeney (1986) use the first minimum of $I(\tau)$ as the optimal time delay for time series prediction. This value of τ also marks the time lag at which $S(t+\tau)$ adds maximal information to what is available from $S(t)$, or, in other words, the redundancy is least (see also Abarbanel, 1996; Kantz & Schreiber, 1997).

The AMI between two measurements X and Y resulting in x_i and y_j respectively can be defined on the basis of equation (3.8) as:

$$I_{XY} = -\sum_i p_X(x_i) \log p_X(x_i) - \sum_j p_Y(y_j) \log p_Y(y_j) + \sum_i \sum_j p_{XY}(x_i, y_j) \log p_{XY}(x_i, y_j) \tag{3.10}$$

where $P_{XY}(x_i,y_j)$ is the joint probability density for X and Y resulting in values x and y, and $P_X(x_i)$ and $P_Y(y_j)$ are the individual probability density for X and Y.

Since $\sum_i p_X(x_i)\log p_X(x_i) = \sum_{i,j} p_{XY}(x_i,y_j)\log p_X(x_i)$ and a similar expression can be used

for $\sum_j p_Y(y_j)\log p_Y(y_j)$ in which case equation (3.10) can be written as:

$$I_{XY} = \sum_{i,j} P_{XY}(x_i,y_j)\log_2\left[\frac{p_{XY}(x_i,y_j)}{p_X(x_i)p_Y(y_j)}\right] \tag{3.11}$$

If measurements X and Y are completely independent, then the AMI I_{AB} is zero.

For a system with continuous probabilities, the Shannon entropy diverges when the partition is refined further and further. Kantz & Schreiber (1997) maintain that the value of the mutual information is independent of the particular choice of histogram, as long as it is fine enough. This is because the ratios of the joint and ordinary probabilities appear as arguments of the logarithm in equation (3.11). This cancels the divergent term and the value of the mutual information approaches a finite limit, which is independent of the partitions used. However it is obvious that, if there is limitation of data, coarse partitions have to be used and this subsequently affects the value of the mutual information.

Correlation Coefficient

In this thesis the AMI and linear correlation functions are used jointly. The correlation coefficient is a measure of association between variables that are ordinal or continuous. The most widely used is the *linear correlation coefficient*. For pairs of quantities (x_i,y_i), $i=1...N$, the linear correlation coefficient r (also called the product-moment correlation coefficient, or *Pearson's r*) is given by the formula:

$$r = \frac{\sum_i (x_i - \bar{x})(y_i - \bar{y})}{\sqrt{\sum_i (x_i - \bar{x})^2}\sqrt{\sum_i (y_i - \bar{y})^2}} \tag{3.12}$$

The value of r lies between -1 and $+1$, inclusive. It takes on a value of 1, termed *complete positive correlation*, when the data points lie on a perfect straight line with positive slope. The value 1 holds independent of the magnitude of the slope. If the data points lie on a perfect straight line with negative slope, y decreasing as x increases, then r takes on a value of -1; which is called *complete negative correlation*. A value of r near zero indicates that x and y are uncorrelated. An advantage of the correlation function as a measure of joint information between two variables is that it has an absolute maximum and minimum.

3.5.   Interpretation of correlation and AMI values

The AMI measure is extensively used in this thesis to detect the presence and degree of relationship between time series data. Here are some of the terms used in the interpretation of results from AMI analysis. Consider X and Y to be time series data. Let X be the preceding (or causative) event and Y be the following (or resultant) event.

If the AMI values between $X(t)$ and $X(t+\tau)$ are computed for different values of lag time τ, the following observations can be made:

High serial correlation: when the AMI at smaller values of τ is comparable to that at $\tau=0$.

Low serial correlation: when the AMI values drop abruptly with increasing τ.

Periodicity: when the AMI value rises periodically at nearly regular time intervals. It has to be noted that the AMI versus lag time curve can indicate periodicity but can resolve the period only to the nearest time interval of the time series.

When the AMI between $X(t)$ and $Y(t+\tau)$ is computed for different lag times τ, the following observations can be made:

Immediate response: when the resultant variable Y responds immediately to changes in the causative variable X regardless of the actual magnitude of the AMI. This is detected when AMI values are high corresponding to $\tau=0$ and decrease with increasing τ.

Delayed response: when the resultant variable Y responds some time later. This is detected when the lag time corresponding to the maximum AMI is non-zero. In this case τ is the average time information takes to propagate from X to Y. Examples of an immediate and a delayed response are illustrated in Figure 3.1.

Strong and weak response: these terms are used when comparing the response of Y to different causative events and are relative measures. If $Max(I(X_1,Y))> Max(I(X_2,Y))$, then it can be said that Y has a stronger response to X_1 than X_2.

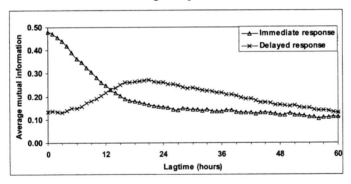

Figure 3.1. Examples of immediate and delayed responses

3.6.   Introducing the weighted mutual information

Using the AMI, the possibility of determining the average separation time between two events described by measures X and Y can be determined. However this time of separation will remain an average. It is possible that the nature of the process connecting X and Y could be such that this time of separation is different depending on, say, the magnitude of X. A typical example of this is flood propagation in rivers where the wave speed is known to vary with the magnitude of the discharge.

In this and other similar problems where the time of separation between two measures of interest depends on the magnitude of either quantity, it is essential to localize the time of separation in order to obtain more specific information on the time dynamics of the process. If the AMI has to be used for this purpose, then it has to be modified. Abebe & Price (2002c) introduced the *Weighted Mutual Information* (WMI) measure by imposing a weighting pattern on the first joint probability term in equation (3.11), which was then applied to determine the relationship between discharge and wave speed in rivers. The purpose of the weight pattern is to give more focus to particular values of the preceding event.

The WMI, designated here as W, for class (or bin) k of the histogram of the preceding event can be defined as:

$$W_{XYk} = \sum_{i,j} w_i p_{XY}(x_i, y_j) \log_2 \left[\frac{p_{XY}(x_i, y_j)}{p_X(y_i) p_Y(y_j)} \right]$$

(3.13)

where w_i is the weight pattern that gives emphasis to class k of the measurement X. If $w_i=1$ for $i=1$ to N_X then $W_{XYk}=I_{XY}$ where N_X is the number of classes (or bins) used in grouping event X.

The weight pattern can be introduced in such a way that it gives full bias to the class interval k and 0 for the rest. That could be done only if the distribution of the available measured data in all the classes is adequate to carry out this analysis. In most hydrological situations that will not be the case since the frequency of extreme events is low and thus not only long but also diverse records would be needed to use the weighted mutual information analysis.

To cope with data scarcity, a weight pattern that allocates a higher value for a particular class of the data and fades away farther in both directions is recommended. At first, a simple and convenient weight pattern that allocates higher weight for class k and linearly fades away in either direction is considered as shown in equation (3.14).

$$w_i = \frac{N_X - |k - i|}{N_X}$$

(3.14)

However, linearly varying weights may not suppress the information from other classes sufficiently and the resulting information might be highly influenced by the adjacent classes. If there is "sufficient" data, it is necessary to concentrate the weight on the particular class k. To maintain a balance between full bias and data scarcity an exponent n is introduced to equation (3.14) resulting in the weight pattern shown in equation (3.15):

$$w_i = \left(\frac{N_X - |k - i|}{N_X} \right)^n$$

(3.15)

where n is an exponent used to manipulate the contrast of the weight pattern.

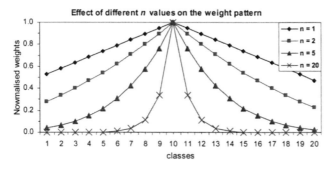

Figure 3.2. Effect of the weight contrast factor n on the weight pattern (the maximum weight for each n is normalized to unity)

The value of the exponent n has the following implications:

if $n=0$ then the WMI and AMI will be equal

if $n=1$ then the weight will fade away linearly from class k.

if $n \to \infty$ then the weight will be exclusive to the class k.

Therefore, the choice of n has to be made depending on the quantity and diversity of the data record available for analysis, in which case expert judgment will be important. Figure 3.2 shows the effect of the weight contrast factor n on the weight pattern on a data set divided into 20 classes. In this case, the class interval under focus, k, is 10. In the case where n is set to 20, the weight pattern is more dominant for class 10 than in cases where lower n values are considered. It has to be noted that the numerical value of W is not important since it might be affected by the weight pattern. However, the lag time corresponding to the maximum value of the WMI is important since it represents the information travel time of a particular range of values of the event X.

3.7.   Principles of uncertainty

3.7.1.   Principle of maximum uncertainty

The principle of maximum uncertainty, also known as the principle of maximum entropy, (Jaynes, 1982) is applicable for reasoning with insufficient information. The principle requires that conclusions resulting from the inference maximize the relevant uncertainty with the available information. In simpler terms, this principle guarantees the full recognition of ignorance while making maximum use of the available information. The principle has been applied in parameter estimation for hydrological models (Singh, 1998) and morphological models (Droste, 1998).

3.7.2.   Principle of minimum uncertainty

The principle of minimum uncertainty states that when faced with alternative solutions that are otherwise equivalent, the best solution is the one involving the minimal amount of uncertainty relevant to the problem. This principle is highly applicable to problems involving simplification. Simplification can be done by eliminating some processes from a model, by aggregating several processes into one, or by breaking down larger processes into appropriate smaller ones (see Klir & Wierman, 1998). If simplification involves eliminating one of the processes in a model, this principle suggests that the process eliminated has to be the one with maximum redundancy.

For instance, if it is intended to simplify a neural network model by cutting some of the input data, the best candidate to be cut is the one that relates the most to one of the other inputs and the least to the output. This principle is applied extensively in this thesis in developing data-driven models. The mutual information measure is used to evaluate the redundancy.

3.7.3.   Principle of uncertainty invariance

The principle of uncertainty invariance proposed by Klir (1990) is based on the measure of uncertainty in various theories and holds that the amount of uncertainty associated with a situation must be preserved no matter what theory (probability, evidence, fuzzy set, etc) is used to conceptualise it. The principle also guarantees a meaningful transformation from one measure to another. Abebe *et al.* (2000a) indicated that fuzzy set and Monte Carlo simulation techniques provide similar information regarding the propagation of uncertainty in a groundwater solute transport model.

3.8.   Case study: flood wave propagation in rivers

The speed of a flood wave in a river is regarded as a very important parameter in flood routing models. In simpler routing models this parameter is considered as a constant. However, a flood propagates from upstream to downstream of a river reach at varying speeds that largely depend on the magnitude of the discharge but can differ from event to event.

In this case study, principles based on information theory are applied to establish the connection between the wave speed and discharge in rivers. The main advantage of the presented approach derives from the fact that it uses only the time series of upstream and downstream water level or discharge data. The method is applied to estimate the travel times of flood waves corresponding to different magnitudes of the discharge using only measured flow time series at the two ends of a reach. It is demonstrated using data from the Wye River in the UK and the Meuse River in The Netherlands.

The material in this case study is adapted from Abebe & Price (2002c).

3.8.1.   Existing approaches

Determination of the relationship between wave speed and discharge is a complex problem mainly because of attenuation of flood peaks and variation in the shape of the cross-section along the reach. If it is assumed that the flood wave does not change its shape along the reach and if the flow is a sole function of the depth, then the speed of the wave is the same as that of a kinematic wave (see Henderson, 1966; Chow et al., 1988) and can be written as

$$c = \frac{dQ}{dA} = \frac{1}{B}\frac{dQ}{dy} \tag{3.16}$$

where c is the wave speed, Q is the flow, A is cross-sectional area, B is the surface width, y is the depth of flow.

Besides its limitations due to the above assumptions, the practical application of equation (3.16) may be limited since it requires cross-section $B(y)$ and rating information $Q(y)$ at all the sections considered. Equations derived from equation (3.16) have been applied by Sriwongsitanon et al. (1998) on the Herbert River in Australia.

For particular flood crests, the speed of a flood wave between two reaches can be calculated as

$$c = \frac{L}{T_P} \tag{3.17}$$

where L is the length of the reach and T_P is the time difference between the upstream and downstream peaks. Equation (3.17) can only be used for flows corresponding to hydrograph peaks available in the record. In addition, it does not take advantage of the information available in the rest of the hydrograph other than the peaks.

Price (1973, 1985) developed a functional form of the relationship between flow and wave speed that takes the attenuation into account:

$$c = \frac{L}{T_P} + \frac{d}{dQ}\left(\frac{L}{T_P}\right)Q^* \tag{3.18}$$

where Q^* is the attenuation of the flood peak. This approach was applied to several British rivers. However, as it is stated in Wong & Laurenson (1983), the correction term for attenuation requires cross-sectional information $B(y)$. The cross-section of a natural river varies along a reach and further assumptions, such as using the average cross-section, have to be made in order to apply (3.18).

3.8.2.   Calculating wave speed using WMI

The WMI can be computed for every class interval k of the preceding event at varying lag times. For a flood routing problem, the preceding event is considered as the upstream hydrograph. The lag time (T_w) at which the WMI corresponding to a class of upstream discharges is a maximum is considered as the average time at which this class of discharges takes to propagate along the river reach. The wave speed for class k of the upstream discharge can then be computed as

$$c_k = \frac{L}{T_{Wk}} \quad for \ k = 1 \ to \ N_X \tag{3.19}$$

where N_X is the number of classes used for the histogram of the upstream flow time series.

In effect, what the WMI method does is to track the time at which information about particular segments of the upstream hydrograph are best related to those of the downstream hydrograph. The wave speed obtained from equation (3.19) corresponds to the average of upstream discharges in class k. However, since the corresponding downstream discharge is different due to attenuation and diffusion, eventually this wave speed has to be mapped to the average of the upstream and corresponding downstream discharges. An important question here is what is the corresponding downstream discharge? One possibility is to assume that the corresponding downstream discharge can be considered as the one falling in the same class index as the upstream flow. The other possibility is to repeat the WMI computation using a weighting pattern that depends on the downstream discharge and compute the corresponding wave speeds so that a mapping can be made between corresponding upstream and downstream discharges. The latter possibility seems more reasonable; however, further research is needed to establish a rigorous guideline for wave speed-discharge computation using the WMI approach.

3.8.3.   River Meuse, The Netherlands

The first test case is done on the Dutch part of the Meuse River on the 91 km reach between locations Borgharen and Venlo. Hourly flow data from 1997 to 2001 was used. A mass balance analysis of the records shows that the average inflow to the reach amounts to 70 m³/s.

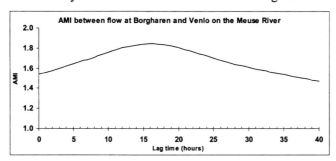

Figure 3.3. AMI between flows at Borgharen and Venlo on the Meuse River

The AMI between the flow data at Borgharen and Venlo is computed up to a lag time of 40 h as shown in Figure 3.3. The maximum AMI corresponds to 16 h. The average wave speed is computed as 91 km/16 h = 1.58 m/s. This wave speed is an average for the whole data used in the analysis. In order to see the extent to which the wave speed varies with the discharge, a series of scatter diagrams of the flows at Borgharen versus Venlo are plotted by varying the time difference in between the two time series (Figure 3.4). It can be observed that different parts of the scatter diagram shrink to the diagonal at different lag times. For instance, discharges between 1500 and 2000 m³/s align more at 28 h of lag time than at lower lag times. The wave speed versus discharge for this reach of the Meuse is computed using weighted mutual information. The results are plotted in Figure 3.5.

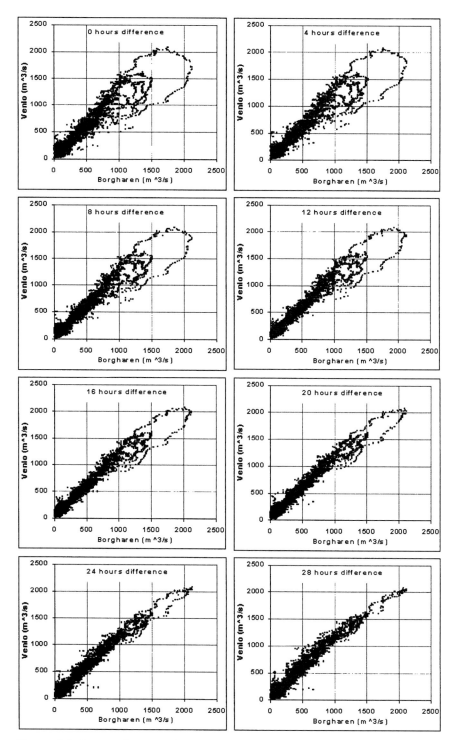

Figure 3.4. Scatter diagrams between discharges at Venlo and Borgharen on the Meuse at varying lag times

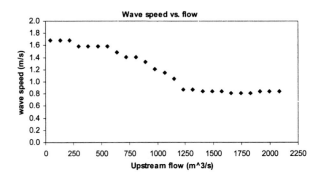

Figure 3.5. Wave speed versus discharge on the Meuse River

3.8.4.   River Wye, UK

The second case study was performed on the 70-km reach between gauging stations Erwood and Belmont on the Wye River in the UK. The quarter hourly discharge recorded at the two measuring stations taken during the flood events that occurred in February 1990 and November 1992 were used.

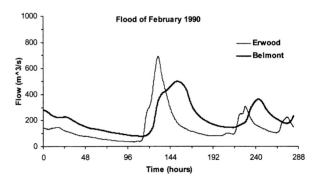

Figure 3.6. Flood of February 1990, River Wye, UK

Figure 3.6 shows the flow records of the 1990 event. From the hydrographs it can be seen that the flow spills to the flood plains when it is in excess of about 400 m^3/s. The baseflow part of the hydrographs shows that there is considerable lateral inflow between the two reaches. The attenuation corresponding to an upstream flow of 650 m^3/s amounts to 200 m^3/s.

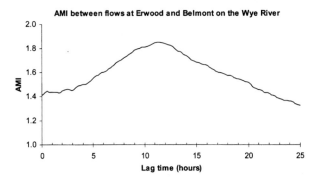

Figure 3.7. AMI between flows at Erwood and Belmont, River Wye, UK

The AMI between the flow at Erwood and Belmont computed using the data from the flood events of 1990 and 1992 is shown in Figure 3.7. The AMI values are computed up to a lag time of 25 h. The maximum AMI corresponds to 11.5 h. The average wave speed can be computed as 70 km/11.5 h = 1.69 m/s.

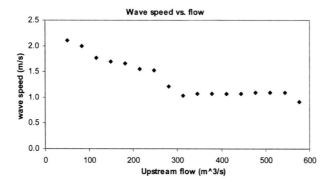

Figure 3.8. Wave speed versus discharge on the Wye River

The wave speed versus discharge obtained using the WMI technique is shown in Figure 3.8. The figure shows a break in the relationship between wave speed and discharge corresponding to discharges between 200 and 300 m^3/s. This computation is done using the hydrographs of only two flood events. For low discharges, care should be taken in interpreting the results since the magnitude of the lateral inflows could be more significant. The scatter diagram is more or less consistent with the theoretical behaviour of the wave speed.

3.8.5.   Conclusions

The results indicate that the weighted mutual information technique presents a promising and viable alternative to calculate/estimate the speed at which flood waves propagate from upstream to downstream of a river. Its application is particularly important when the only available data is the discharge or water level time series at two ends of a river reach. In its present form, the weighted mutual information approach can be used in river reaches that have flow or stage measuring stations at both ends. Although its effect on the method is not yet properly established, the lateral inflow has to be minimal compared to the main flow.

The time series of data used for the Meuse is continuous and longer than the one used for the Wye. The corresponding result is also a smoother relationship between discharge and wave speed than the result for the Wye River. The wave speeds corresponding to very low discharges (<100 m^3/s) the Wye River could not be computed since there are no records of such magnitude in the data.

Further investigation is needed to establish the applicability of this information theory-based approach in a wide variety of river flow conditions particularly at the transition between in-bank and floodplain flows. Also it is necessary to compare the results obtained from this approach to those obtained by conventional approaches. Perhaps, the best way for such a comparison is to use synthetic data generated under controlled conditions.

3.9.   Information theory in data-driven modelling

The AMI measure can be applied in the processes of developing data-driven models. Since AMI can be used to measure the strength of the relationship between two data series, it can be used to determine which data are the best related input data for a data-driven model that is intended to predict some output. It can also be used to measure the relationship between different input data of a data-driven model. If it is intended to simplify the model by eliminating some of the input data, AMI measures can be used as a measure of redundancy between input data, and the most redundant data can be excluded.

The weighted mutual information measure has even more enhanced features that can be used in data-driven modelling, in that it helps to identify the time dynamics between two data sets. Since the data might be best related with each other at different lag times depending on the magnitude of one of the data series, WMI can be used to mark the separation line between shifts in the time dynamics of the relationship.

For example, in the case of the Meuse River, the WMI analysis showed that for low flows the information travel time is 16 h whereas for high flows it is 30 h. If a data-driven model that relates upstream and downstream flows was to be developed, this implies that to establish a good input-output relationship, it would be better to develop two separate models one for high flows and another for low flows. The region in between can then be filled with some form of interpolation. This is because the source of maximum information from the upstream measurement location is different depending on the magnitude of the flow. The use of two separate models for high and low flows is explored by See & Openshaw (1999) by applying multiple neural network models to the Ouse River, UK. The use of information measuring techniques would further help in overseeing the process of selecting input data in data-driven modelling.

In this study, principles based on information theory are applied extensively in analysing the relationship between data, studying the time dynamics involved and preparing data for the development of data-driven models.

CHAPTER 4. ARTIFICIAL INTELLIGENT APPROACHES

The term artificial intelligence refers to the use of machines to do tasks that would otherwise need human intelligence to accomplish them (Boden, 1977). In recent decades, advances in computing have given a great momentum to the development and application of artificial intelligent systems. There are various techniques grouped under artificial intelligence (see, for example, Negnevsky, 2002 for a comprehensive review). This chapter gives an overview of artificial neural network and fuzzy logic based techniques. Both are extensively applied in the following chapters in modelling the relationship between data.

4.1. Artificial neural networks

Artificial Neural Networks (ANNs) are interconnected structures composed of a number of units known as artificial neurons characterized by input/output nodes and activation functions, which work at a sub-symbolic level to represent a relationship through connecting weights. ANNs are nowadays regarded as universal function approximators (Cybenko, 1989; Hornik *et al.* 1989) due to their ability to represent the relationship between data as input and output with reliable accuracy. They have been applied successfully in control, classification, and modelling dynamical systems and time series forecasting. ANN models are being utilized as a viable alternative to other modelling approaches. There are various architectures of neural networks each suited to particular applications.

In recent years, the ANN modelling approach has been a mainstream area in hydrodynamics, river flow forecasting and rainfall-runoff modelling. For instance, Dibike & Solomatine (1999) showed that the simulation results of a MIKE 11 model of the Apure River in Venezuela can be mimicked with an ANN model to reduce the running time for optimisation purposes. Tingsanchali & Manusthiparom (2001) demonstrated the applicability of neural networks to forecast floods 3 hours ahead on the tide-affected part of Chao Phraya River in Thailand. See & Openshaw (1999) and Lekkas *et al.* (2001) also used ANN routing models for real-time forecasting. Additional applications of ANN models are referenced in other chapters of the thesis.

In this thesis, ANN models are applied in most of the case studies. This section provides an overview of commonly used types of ANNs and the important issues surrounding their application that are referenced in other parts of the thesis.

4.1.1. Artificial neurons

The model of the artificial neuron was first proposed by McCulloch & Pitts (1943) as a binary device using binary inputs, binary output and a fixed activation threshold. An artificial neuron (Figure 4.1) consists of the following properties:

The *inputs* $x_1, x_2, \dots x_n$ that are associated with the connection weights $w_1, w_2, \dots w_n$.

A *function f*, to calculate the aggregated net input signal to the neuron $u = f(\mathbf{x}, \mathbf{w})$, where $\mathbf{x}$ and $\mathbf{w}$ are the input and weight vectors. The function f is usually the weighted summation of the inputs: $u = \sum x_i w_i$ for $i = 1 \dots n$

An *activation function s* that calculates the activation level of the neuron: $a = s(u)$

An *output function g* that calculates the output signal of the value passed to the output of the neuron: $o = g(a)$. Usually the output signal is assumed to be equal to the activation level of the neuron: $o = a$.

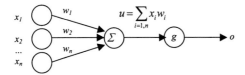

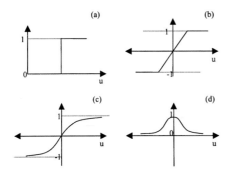

Figure 4.1. An artificial neuron

The main difference between the neurons in common use lies in the type of the activation function. Depending on what the neuron is intended to do, one of the activation functions shown in Figure 4.2 can be used directly or in a modified form. For instance, a binary activation function (Figure 4.2(a)) can be used when the neuron is intended to classify the signals based on a threshold value. A linear activation function (Figure 4.2(b)) is used to map an input between two thresholds into a linear output. Sigmoid and Gaussian functions (Figure 4.2(c) and (d)) provide non-linear and continuous outputs corresponding to an input signal.

Figure 4.2. Commonly used activation functions: (a) binary function, (b) linear function, (c) sigmoid function, and (d) Gaussian function

4.1.2.   Training artificial neurons: the perceptron

Rosenblatt (1958) introduced an algorithm that provided the first procedure for training a simple artificial neuron known as *the perceptron*. Consisting of only a single neuron with adjustable weights and a threshold, the perceptron is the simplest form of a neural network. The perceptron essentially divides into two the hyperspace of inputs by a hyperplane thus providing a binary output as

$$y = Sign\left(\sum_{i=1,n} x_i w_i - \theta\right) \tag{4.1}$$

where **x** and **w** are the input and weight vectors, y is the output and θ is the threshold.

The perceptron is trained by making small adjustments to the weights starting from random weights. The adjustment is made iteratively based on the algebraic sign of the error between the expected and actual output signal until the error is low enough. The main limitation of the perceptron is that it resolves only linearly separable patterns regardless of the activation function used (Shynk, 1990). A single neuron might perform simple transformation functions. Its power in real life application, however, lies on its connectionist character, which takes the artificial neuron into the multi-layer neural network.

4.1.3. The multi-layer perceptron

The multi-layer perceptron (MLP) is the most commonly used of the multi-layer neural networks. The MLP consists of an input layer, an output layer, and at least one intermediate layer. The neurons of one layer are connected to the neurons of the next layer. The first layer is the *input layer* as it receives the input signal. The last layer is the *output layer* since its outputs are considered as outputs of the network. The intermediate layers are collectively known as *hidden layers* since the neurons in these layers are not connected to the outside world. In most practical applications, only a single hidden layer is used. This is partly to avoid over-parameterisation and partly since it has been proven that one hidden layer works as good as multiple ones (see Cybenko, 1989).

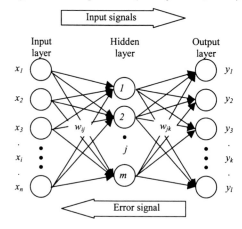

Figure 4.3. A three layer MLP network

The model of a three-layer MLP is shown in Figure 4.3. The MLP has a *feed forward* architecture meaning that there are no connections from the output back to the input neurons. During operation, the weighted sums of the input vector pass through the activation function to result in the values of the hidden layer. The weighted sums of these values are again passed through the activation function of the output layer and result in the network output.

4.1.4. Training MLP networks: back propagation

Following the conception of the neuron, the development of ANNs went into a dormant state mainly due to the then lack of a robust methodology to determine the connecting weights. The process of determining the weights of an ANN from example data is known as *training*. The training data consist of a series of input and output vectors whose relationship the ANN is supposed to learn. For a given network structure, also known as *topology*, the connection weights are the parameters that enable the network to learn the relationship between the training data and adapt to the process that generated the data. Right now there are various algorithms to train MLP networks (see Haykin, 1994). The most popular training algorithm for MLP networks is the *back propagation* (BP) algorithm (Rumelhart *et al.* 1986) also known as the error back propagation algorithm.

As its name implies the BP algorithm works with a series of feed forward and back propagation iterations starting from a random set of connecting weights. At the feed forward stage, the input vectors of all the training examples are fed to the network and the output vectors are calculated. The performance of the network is evaluated using an error function that is based on the target and network outputs. At the back propagation stage, the error is propagated back to calculate the adjustments needed to the network weights. The iteration continues until the stopping criteria are met. The trained ANN is then tested on an independent data set known as *verification data*. Various ways of verifying trained networks are discussed in Kasabov (1996).

Usually, the input and output of the training data are scaled to a regular interval. This procedure, often termed *standardisation*, helps to prevent computational overflow as a result of data with different orders of magnitude.

4.1.5.   Radial basis function networks

The other most popular type of ANN also used in parts of this thesis is the *radial basis function* (RBF) network (see Moody & Darken, 1989). Consisting of three layers (see Figure 4.4), the first layer has *n* inputs that are connected to the neurons in the second (hidden) layer. The activation function of the nodes in the hidden layer is an RBF. This has a radially symmetrical basis function, such as the Gaussian function, which can be written as:

$$f(x) = e^{-\frac{(x-\mu)^2}{2\sigma^2}} \qquad\qquad (4.2)$$

where μ and σ are the mean and standard deviation of the input x.

For a particular hidden node i, its RBF_i is centred at a cluster centre c_i in the *n*-dimensional input space. The cluster centre c_i is represented by the vector $(w_{1i}, ..., w_{ni})$ of connection weights between the *n* input nodes and the hidden node i. The standard deviation for this node defines the range for the RBF_i. The RBF is non-monotonic in contrast to the sigmoid function. The hidden layer is connected to the output layer. The output nodes in the third layer perform a summation after a linear activation.

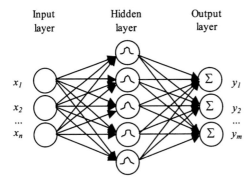

Figure 4.4. RBF networks

Training an RBF network consists of two phases: adjusting the RBF of the hidden neurons by a clustering method, and applying a gradient descent method to adjust the second layer of connection weights. During application, the functions RBF_i are used to evaluate how close an input vector **X** is to the centres **C**$_i$. The distance values are then propagated to the output layer. As in MLP networks, the actual input signals are usually standardised before entering the network and the outputs are scaled back to their actual magnitude.

RBF networks have the advantages of faster learning for networks with a limited number of input nodes, better generalization, and the absence of local optimum problem during training

(Kasabov, 1996). However, a disadvantage is that for larger number of input nodes, the number of hidden nodes grows quickly.

4.1.6.   Supervised and unsupervised learning

In supervised learning, the network performance is judged by comparing the actual network output to a desired output or target for the corresponding input. The training algorithms for MLP and RBF networks discussed above are intended to adjust the network weights to achieve network outputs that are closer to the target. This form of learning is called supervised learning since it is based on training data composed of input and desired output.

In contrast, unsupervised learning algorithms try to find classes or clusters in data. There are several approaches to clustering algorithms. For example, one might attempt to find clusters of data points with all data within a cluster being similar to each other. Another approach would be to find groups where the emphasis is on the groups being distinct, rather than the points within the group being similar. These can be accomplished by minimizing different forms of objective functions. The self-organizing feature maps (SOFM), also known as Kohonen maps (Kohonen, 1990), use unsupervised learning, meaning the training is done using only input vectors. Unsupervised learning is useful in discovering new facts, similarities and categories from data.

4.1.7.   Practical problems with ANN models

Over parameterisation

An important issue in choosing an ANN model for a given problem is selecting the level of structural complexity that suits the problem and available data. If the model contains too many parameters, it will approximate not only the data but also the noise in the data. On the other hand, a model that contains too few parameters will not be flexible enough to approximate important features in the data. Thus the modeller should strike a balance between the size of the network on one side, and the complexity of the relationship and the availability of training data on the other. The number of nodes in the hidden layer in relation to those in the input and output layers specifies the flexibility of an ANN model. There is no known rigorous way to fix the number of hidden nodes, however, there are guidelines to determine the optimal size of networks. These guidelines are generally based on relating the number of training examples to the number of weights trained, the number of hidden nodes to the number of input nodes, etc (see, for example, Hecht-Nielson, 1991). A comprehensive study on various optimisation techniques to determine the neural network structure is done in Zijderveld (2003).

A danger associated with an oversized network is that it demands too much data to train its weights. The network might as well have a structure much more complex than what the actual process or relationship between the data demands. If the amount of data is not sufficient in comparison to the number of connecting units in the ANN model, there will be a problem of indeterminate parameters.

Overfitting

Over fitting refers to the undesirable effect of training an ANN model that results in the fitting of the network to noise in the training data. Overfitting reduces the generalizing ability of the ANN model during testing and application.

There are ways to minimize overfitting of ANN models to noise in the training data. A logical start to overcome overfitting is the use of a network structure that is optimised not only to the complexity of process under consideration but also to the amount and diversity of data available for training.

In the presence of sufficient data, the best way to overcome overfitting is early stopping. Early stopping involves the use of a third data set in addition to the training and verification data, which is known as cross-validation data set. The training is intended to minimize the error on the training data. Meanwhile, the cross-validation error is monitored at intervals. The training process is stopped when the error in the cross-validation data begins to rise even when the training error is still decreasing.

However, there might not be the luxury of ample exemplar data to be split into three and have a cross-validation set. Under such circumstances, the training process has to be limited to the training set. A commonly used way to overcome overfitting is to do the training on the basis of minimizing a function composed of the training error and a cost function. The cost function is used to penalize the search towards large weights. This approach is known as *weight regularization*.

For instance, the so-called Akaike Information Criterion (AIC) (after Akaike, 1974) uses a measure based on entropy as shown in equation (4.3).

$$AIC(K_a) = -2\log(L) + 2K_a \tag{4.3}$$

where L is the maximum likelihood of the model parameters and K_a is the number of adjustable parameters in the model, weights in this case. In equation (4.3), the log-likelihood term tends to regularise the weights whereas the second term penalises increased number of weights.

Bishop (1995) proposed used a cost function composed of an error function and a term that penalizes the search towards large network weights. If there are N training examples and knowing the true output, $y(\mathbf{x})$, the cost function C can be written as:

$$C = \frac{1}{N} \sum_{i=1,N} \left(y_i - f(\mathbf{x}_i) \right)^2 + \lambda \sum_{j=1,M} w_j^2 \tag{4.4}$$

where $f(\mathbf{x})$ is the network output for input $\mathbf{x}$, $y(\mathbf{x})$ is the true output, N is the number of training examples, λ is a weight regularization factor and w_j, $j=1\ldots M$ are the connecting weights in the network. In equation (4.4), the first term is the MSE whereas the second is the penalty term.

Extrapolation

Extrapolation is the prediction of values of unknown data points by projecting a function beyond the range of known data points. Application of ANN models to data outside the range of the data used for training poses a great deal of uncertainty. The main reason is that most activation functions have a threshold that limits the corresponding output signal. Some activation functions such as the sigmoid function might restrict extrapolation since the response is asymptotic and practically insensitive for input data exceeding some value.

The inability of neural networks to extrapolate beyond the range of training data is widely acknowledged (see Minns & Hall, 1996) and there are very few examples of neural network configurations with improved extrapolation ability reported in the literature (Imrie *et al.*, 2000). This is important in forecasting applications, since the probability of events exceeding the range covered by the training data may be low but still remains different from zero. A possible approach the extrapolation problem is by altering the standardization range of the data. For instance, standardizing the data to a narrower portion of the activation function such as [0.2,0.8] for the output of the activation function being in [0,1] enables the ANN to extrapolate (see Maier & Dandy, 2000; Varoonchoticul *et al.* 2002). Still, the real danger is that the response of the actual process might not be just an extrapolation of the relationship encapsulated in the trained network, which might be over-learning instead of extrapolating (Minns, 1998). Even if the ANN outputs exceed the training range, there could be a significant divergence from how the observed data would have responded. At this point, knowledge of the physical process that generated the data is essential.

4.2.   Fuzzy systems

Fuzzy logic, also known as fuzzy set theory, emerged after the pioneering work of Zadeh (1965) as a more general form of logic that can handle the concept of partial truth. In fuzzy logic, truth can take intermediate values between "completely false" and "completely true". Since then it is serving as a modelling methodology that allows a better way of handling imprecise information. The application of fuzzy systems has increased in water-based problems. One reason for this is the flexibility of fuzzy sets that allows an efficient means of knowledge representation.

A study by Carpa *et al.* (1994) applied fuzzy set theory to drought classification. They used fuzzy clustering techniques to identify areas with similar meteorological characteristics. Pesti *et al.* (1994) proposed a methodology for predicting regional droughts from large-scale atmospheric patterns. They used a fuzzy rule-based model (FRBM) to predict the so-called Palmer's Drought Severity Index of the New Mexico area based on atmospheric circulation patterns of the United States. Bárdossy *et al.* (1995) modelled the movement of water in the unsaturated zone using FRBM. Data generated by a numerical solution of Richard's equation was used to formulate the rules of an FRBM, which essentially replicate a complex mathematical model. Bárdossy & Duckstein (1995) also used fuzzy systems to model the daily water demand time series in the Ruhr basin, Germany, and used fuzzy rules to predict future water demand. Fuzzy set theory is also applied in decision-making problems. Fontane *et al.* (1997) posed a multi-objective reservoir operation problem using linguistically described operational goals and fuzzy constraints. Their study included conducting interviews of both decision-makers and representatives of decision-influence groups to develop measures of multiple fuzzy objectives in terms of reservoir releases or storage. Abebe *et al.* (2000b) applied FRBM on a problem of reconstructing missing precipitation records based on data from adjacent recording locations.

The two most widely used forms of its application are fuzzy rule-based modelling (FRBM) and the fuzzy alpha-cut technique. This section provides an overview of the basics of fuzzy set theory. It also discusses the working principles of a fuzzy rule-based system along with the algorithms commonly used to extract fuzzy rules from data. A case study on its application in analysing the propagation of uncertainty in models is also described. The FRBM approach is applied in the later parts of this thesis.

4.2.1.   Fuzzy sets and membership functions

A fuzzy set is a set without clear boundaries. Contrary to crisp sets, an element of a fuzzy set can have partial membership. In mathematical terms, a fuzzy set A on a universe U is such that, for any $u \in U$, there is a corresponding real number $\mu_A(u) \in [0, 1]$ to u, where $\mu_A(u)$ is the degree to which u belongs to A. This can be shown with following mapping:

$$\mu_A : U \to [0,1], \quad u \to \mu_A(u) \tag{4.5}$$

The mapping function is called the *membership function* of A.

If the range [0,1] is replaced by just two values zero and one, then A reduces to a crisp set.

$$A = \{u \in U \mid \mu_A(u) = 1\} \tag{4.6}$$

Therefore, one can say that crisp sets are special cases of fuzzy sets or fuzzy sets are generalized forms of ordinary sets.

A membership function can be of any shape depending on the type of fuzzy set it belongs. For computational implementation, membership functions can be represented by piecewise linear equations, by Left-Right (L-R) curves (Bárdossy & Duckstein 1995), or the composition of other functions. If L-R curves are used, the membership function will have the structure shown in the equation below:

$$\mu_A(x) = \begin{cases} L(x) & if \quad \alpha^- < x < \alpha^1 \\ 1 & if \quad x = \alpha^1 \\ R(x) & if \quad \alpha^1 < x < \alpha^+ \\ 0 & else \end{cases} \tag{4.7}$$

The range $[\alpha^-, \alpha^+]$ is called the *support* of the fuzzy set whereas α^1 is its *kernel*. When the kernel of the membership function reduces to a point and the L-and R-functions become linear, the function reduces to the most widely used triangular membership function as shown in Figure 4.5.

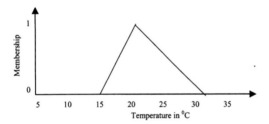

Figure 4.5. Example membership function of "comfortable" room temperatures

An important feature that comes with fuzzy sets is that they can be associated with linguistic variables. Zadeh (1975) has suggested a mathematical transformation for the most commonly used linguistic modifiers such as "very" "some what" and "more or less" which can be used to maintain a closer tie with natural language.

4.2.2.   Fuzzy rules

Fuzzy rules are IF-THEN- statements that have the general form "IF A THEN B" where A and B are (collections of) propositions containing linguistic variables. A is called the *premise* and B is the *consequence* of the rule. In a more explicit form, if there are *I* rules each with *K* premises in a system, the ith rule can have the following form:

$$IF \ a_1 \ is \ A_{i,1} \ \Theta \ a_2 \ is \ A_{i,2} \ \Theta \ldots \Theta \ a_k \ is \ A_{i,k} \quad THEN \ B_i \tag{4.8}$$

In the above equation the a_k represent the crisp inputs to the rule and Θ is an operator such as AND or OR.

An example of a fuzzy rule is:

IF expected flood is HIGH AND reservoir level is MEDIUM, THEN water release is HIGH.

4.2.3.   Fuzzy rule-based models

A fuzzy rule-based model (FRBM) is essentially a model that represents the relationship between causative and resultant data using a collection of fuzzy rules. In this context, Kosko

(1993) describes fuzzy rules as "patches" of local models overlapped throughout the parameter space, using a sort of interpolation at a lower level to represent patterns in complex non-linear relationships.

An FRBM contains membership functions of fuzzy sets constructed over the range of its input data. Since membership functions generally overlap each other, so do the rules constructed from them. The system of rules in an FRMB is often represented in the form of a *fuzzy associative memory* (FAM) matrix (Kosko, 1992). A FAM matrix is an N dimensional table where each dimension represents a specific input with size equal to the number of fuzzy sets used to describe that input. So, for example, if the input vector is comprised of 2 elements with each input described by 3 fuzzy sets the matrix would have $3^2=9$ entries. Each FAM entry is used to store the consequent of one rule.

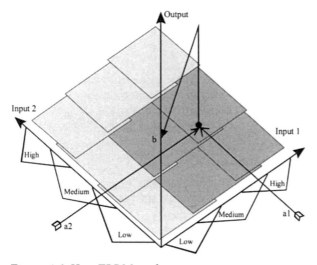

Figure 4.6. How FRBM works

Figure 4.6 shows an FRBM with two inputs (*a1* and *a2*) and one output (*b*). When a vector of input data is fed into the model, memberships of the corresponding input fuzzy sets are determined. This step is known as *fuzzification*. For instance, *a1* belongs to HIGH and MEDIUM on *input 1* whereas *a2* belongs to LOW and MEDIUM on *input 2*. This activates four of the nine overlapping rules, which are shown filled with darker tiles.

Inference and degree of fulfilment

In traditional rule-based systems where crisp rules are used, a rule has a binary function of fulfilment of its premises: it is satisfied or not satisfied. In fuzzy systems, however, each input belongs to the corresponding fuzzy set to varying degrees. Consequently, the rules are satisfied to varying degrees measured on the continuous scale [0,1] depending on the conditions of the rule. The extent to which the premise of a fuzzy rule is satisfied is called the *degree of fulfilment* (DOF). The DOF determines to what extent each rule is applied. The rules involving non-zero degrees of fulfilment are activated (as shown with the darker tiles in Figure 4.6). The DOF of each rule is transferred to the corresponding consequent of the rule. This process is known as *inference*. There are two well-known inference mechanisms, one proposed by Mamdani & Assilian (1975) and the other by Sugeno (1985). The basic difference between the two inference mechanisms lies in the fact that the former works with fuzzy consequences whereas the latter uses singletons (crisp values) as consequents

(Negnevitsky, 2002). In both cases, the consequences are eventually combined depending on the type of the problem.

The DOF of the rules can be determined using one of two ways: minimum and product inference. For premises of a rule connected by the AND operator, the DOF according to the product inference is:

$$v_i = DOF(A_i) = \prod_{k=1}^{K} \mu_{A_{i,k}}(a_k) \tag{4.9}$$

In the product inference, all memberships in the premises of the rule are used to determine the DOF. Therefore, it is possible to say that the rules respond to the slightest change in the degree of truth contained in each premise. This makes it suitable to model systems with a numerically continuous response.

Using the minimum inference, the DOF can be computed as:

$$v_i = DOF(A_i) = \underset{k=1..K}{Min}\left(\mu_{A_{i,k}}(a_k)\right) \tag{4.10}$$

In the minimum inference, the determining factor for the DOF in case of premises connected by the AND operator is the premise with the lowest membership among the rule premises, which makes it a winner-take-all approach.. Hence; it is not sensitive to changes in other premises of the rule. This makes it suitable to model systems with a discrete response such as classification problems.

Combination of rule responses and defuzzification

If only one rule is activated at a time, that rule will be executed. More interesting is what happens when multiple rules are activated, in which case the inference engine applies the respective DOFs of each rule to the corresponding consequence. There are several ways to combine the responses of multiple rules. Their consecutive consequences can be combined to obtain a fuzzy or crisp output (see Zadeh, 1983; Dubois & Prade; 1991; Bárdossy & Duckstein, 1995; Kosko 1993). For instance, in the *weighted sum combination*, the DOF of each rule is used as a weight leading to a number of scaled output membership functions as shown in Figure 4.7. If there are I rules each having a response fuzzy set B_i with a DOF of v_i, the combined membership function using weighted sum combination can be written as shown in equation (4.11).

$$\mu_B(x) = \frac{\sum_{i=1}^{I} v_i \mu_{B_i}(x)}{Max_u \sum_{i=1}^{I} v_i \mu_{B_i}(u)} \tag{4.11}$$

The weighted sum combination is illustrated in Figure 4.7. If four of the nine rules shown in the form of tiles in Figure 4.6 are activated, the rule responses are combined to give those shown with thick lines in Figure 4.7(b). The fuzzy inference engine ends up with a fuzzy set that is a subset of the universe of the output variable and has a rather complicated shape as shown in Figure 4.7(b). The process hereafter depends on the intended type of output.

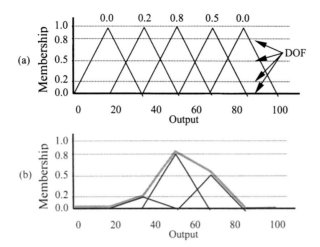

Figure 4.7. Weighted sum combination of rule responses (a) before combination: the DOFs are shown on top of each membership function; and (b) after combination: the solid lines indicate output membership functions scaled with the corresponding DOF, the dim line indicates the sum of all membership functions.

The process of mapping from a fuzzy output or outputs B_i to a crisp output b is called *defuzzification*. If the FRBM is intended for classification, or if discrete outputs are required, the response of the rule with maximum DOF may be taken, which is known as *maximum defuzzification*. If a continuous numerical output is expected, the fuzzy output has to be defuzzified. The most widely used way to defuzzify systems with a continuous numerical response is to take the centroid of the fuzzy response, which is known as *centroid defuzzification*.

4.2.4. Training fuzzy rules

The major task in using FRBM is the formulation of the rules. In simpler systems, the rules, which are the building blocks of a FRBM, could be obtained from expert knowledge. In complex systems and for applications that require numerical accuracy, human expert knowledge may not have the necessary precision. It is vital to use intelligent systems that can configure their own working rules. In such cases, the best way to formulate fuzzy rules is to use data that contain the historical relationship between the premises and the consequences of the fuzzy rules. Analogous to neural networks, the process of formulating fuzzy rules from historical data is known as training. An FRBM that is capable of establishing its working rules is known as adaptive fuzzy rule-based model.

A typical training data set (T) consists of a set of S vectors each of which is composed of K components of the input vector a and output value b as shown in equation (4.12).

$$T = \{(a_1(s), \ldots, a_K(s), b(s)); \ s = 1, \ldots, S\} \tag{4.12}$$

There are several ways of training fuzzy rule based model established in the literature, most of which work with a predetermined rule structure and training data. Some of the training algorithms are:

❑ Weighted counting algorithm

❑ Least squares training algorithm

- ❑ Genetic algorithm

- ❑ Clustering techniques

Weighted counting algorithm

This algorithm, used by Bárdossy ·& Duckstein (1995), works with a known rule structure and is oriented towards the construction of output membership functions corresponding to each rule. It uses the subset of the training set that satisfies the premises of a rule at least to a DOF of ε, (known as the threshold) to construct the shape of the corresponding consequence. It is accomplished by the following steps:

Define $(\alpha_{i,k}^{-}, \alpha_{i,k}^{+})$ of the rule argument $A_{i,k}$.

The $A_{i,k}$ is assumed to be a triangular fuzzy number $(\alpha_{i,k}^{-}, \alpha_{i,k}^{1}, \alpha_{i,k}^{+})_T$ where $\alpha_{i,k}^{1}$ is the mean of all possible $a_k(s)$ values which fulfil at least partially the ith rule:

$$\alpha_{i,k}^{1} = \frac{1}{N_i} \sum_{s \in R_i} a_k(s) \tag{4.13}$$

Calculate the DOF $v_i(s)$ for each premise vector $(a_i(s), \dots, a_k(s))$ corresponding to the training set T and each rule i whose premises were determined above.

Select a threshold $\varepsilon > 0$ such that only responses with $DOF \geq \varepsilon$ are considered in the construction of the rule response. Then the corresponding response is the fuzzy number $(\beta_i^{-}, \beta_i^{1}, \beta_i^{+})_T$ where:

$$\beta_i^{+} = \underset{v_i(s) > \varepsilon}{Max}\, b(s)$$

$$\beta_i^{1} = \frac{\sum_{v_i(s) > \varepsilon} v_i(s) b(s)}{\sum_{v_i(s) > \varepsilon} v_i(s)} \tag{4.14}$$

$$\beta_i^{-} = \underset{v_i(s) > \varepsilon}{Min}\, b(s)$$

Least squares training algorithm

The least square training algorithm is also proposed by Bárdossy & Duckstein (1995). It is oriented towards the construction of output membership functions corresponding to the FAM matrix entries. As its name implies the algorithm minimizes the sum of the square error between the modelled and true values. This method has been applied to model the movement of water in the unsaturated zone and a good performance was reported (Bárdossy *et al.* 1995).

The algorithm works as follows:

Consider the training data T described in equation (4.12). The sum of the squared error resulting from the use of the rule system R can be written as shown in equation (4.15).

$$\sum_s \left[R(a_1(s), \dots, a_K(s)) - b(s) \right]^2 \tag{4.15}$$

As the left-hand side of the rules is supposed to be known, the DOF, $v_i(s)$, corresponding to $a_i(s), \dots, a_k(s)$ can be calculated from each rule i of the total I rules. Then the rule response can be written as in equation (4.16), where $M(B_i)$ is the centre of mass of the area under the output membership function B_i.

$$R(a_1(s),...,a_K(s)) = \frac{\sum_{i=1}^{I} v_i(s)M(B_i)}{\sum_{i=1}^{I} v_i(s)} \qquad (4.16)$$

The objective function to be minimized is the one shown in equation (4.17).

$$\sum_s \left(\frac{\sum_{i=1}^{I} v_i(s)M(B_i)}{\sum_{i=1}^{I} v_i(s)} - b(s) \right)^2 \qquad (4.17)$$

Differentiating with respect to the unknown $M(B_j)$ for every index j gives

$$\sum_s \left(\frac{\sum_{i=1}^{I} v_i(s)M(B_i)}{\sum_{i=1}^{I} v_i(s)} - b(s) \right) \frac{v_j(s)}{\sum_{i=1}^{I} v_i(s)} = 0 \qquad (4.18)$$

Rearrangement of the terms gives a system of linear equations with I unknowns ($M(B_1)$ ... $M(B_I)$) and I equations as shown in equation (4.19).

$$\sum_s \left(\frac{\sum_{i=1}^{I} v_i(s)v_j(s)M(B_i)}{\left(\sum_{i=1}^{I} v_i(s)\right)^2} \right) = \sum_s \frac{v_j(s)b(s)}{\sum_{i=1}^{I} v_i(s)} \qquad (4.19)$$

Abebe *et al.* (2000b) applied this training approach on a problem of reconstructing missing precipitation records based on data from adjacent recording locations. After the initial training is completed, they applied a local search to select the best shape of input membership functions from four predefined membership functions, *viz*, triangular, bell-shaped, dome-shaped and inverted-cycloid functions (Figure 4.8). The local search was vital as it allows the fuzzy model to be fine tuned depending on the non-linearity existing in the relationship between the data. While having high prediction accuracy, the least squares training algorithm might not yield output fuzzy sets that can be mapped linguistically.

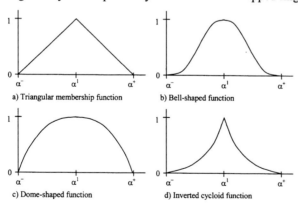

Figure 4.8. Different types of membership functions (adapted from Abebe et al., 2000b)

Genetic algorithms

Genetic algorithms (Goldberg, 1989) have been used to generate fuzzy rules from training data (see for example Ishibuchi *et al.*, 1992; Ishibuchi *et al.*, 1995). When a genetic algorithm is used in training the rules of a FRBM, the hybrid technique is generally known as a geno-fuzzy system.

The way in which genetic algorithms are used with a FRBM can differ. For instance, Herrera *et al.* (1995) used a real coded genetic algorithm where each chromosome represents a fuzzy rule, and then performed an iterative process to obtain a set of rules covering the set of examples. Another approach is to construct a number of membership functions on the output domain. The training process is then posed as a combinatorial optimisation problem that is intended to map input fuzzy sets to output fuzzy sets. Genetic algorithm is used to perform the optimisation after defining an appropriate payoff function (see Welstead, 1994; also Bäch & Kursawe, 1995). Abebe & Price (2003a) applied geno-fuzzy systems to characterize the prediction uncertainty of a model in the form of fuzzy rules. They used a payoff function, which is the inverse of the *RMSE* between the output of the FRBM and training set. In order to use genetic algorithm for training fuzzy rules, at least three conditions need to be satisfied:

(1) The problem has to be posed in such a way that there is a population of possible solutions that can be encoded in the form of bit-strings (known as chromosomes). This is assuming the use of a binary-coded genetic algorithm in which the decoded bit-strings correspond to indices of output membership functions.

(2) There must be a mapping function that transforms a chromosome into the corresponding rule base (phenotype) and back.

(3) There must also be a fitness function that serves as a means of evaluating individual solutions. This function is used to favour the search towards feasible solutions.

The training can start by partitioning the range of input and output of the training set into a number of desired fuzzy sets that can be linguistically described, such as LOW, MEDIUM and HIGH. Their membership functions are also defined. With the number of fuzzy sets covering each input space known, the number of rules is also known. For instance, if there are two inputs with m and n fuzzy sets, the number of rules is (m x n). This essentially defines the size of the FAM matrix. Also at this stage the IF part (premise) of all the rules is known. The remaining task is to assign the proper output fuzzy set to the THEN part (consequence) of each rule. Since this is a discrete optimisation problem, genetic algorithm can be used to find the best output fuzzy set to each rule. The chromosome of each individual contains a complete list of indices representing an output fuzzy set corresponding to each rule. Computing the fitness of each individual involves evaluating the FRBM on the training data. The fitness can be defined as a function of the error between the defuzzified (numerical) output of the FRBM and the corresponding values in the training set. The search proceeds as follows:

(1) Generate a population of potential solutions.

(2) Calculate the fitness of each individual in the population.

(3) IF stopping criteria is met THEN stop, ELSE continue.

(4) Select potential mates in the population.

(5) Mutate individuals.

(6) Crossover to obtain new offsprings.

(7) Put new offsprings into the population and go to (2).

The encoding-decoding process involves a mapping between bit strings and indices of membership functions of the output. The decoding process is illustrated in Figure 4.9 for a fuzzy system with nine rules. For instance, assume that 5 membership functions are constructed on the output with names *very low* (VL), *low* (LO), *medium* (ME), *high* (HI) and *very high* (VH) which are also indexed respectively as 1, 2, 3, 4 and 5. The figure shows that the second rule has a binary code of 101. This can be translated to a decimal code of 5 and represents the output membership function very high (VH). Therefore the consequence part of the second rule is *very high* according to this individual.

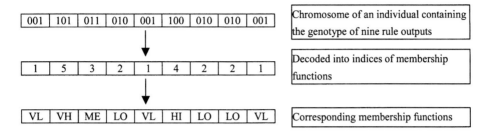

| 001 | 101 | 011 | 010 | 001 | 100 | 010 | 010 | 001 | Chromosome of an individual containing the genotype of nine rule outputs |

| 1 | 5 | 3 | 2 | 1 | 4 | 2 | 2 | 1 | Decoded into indices of membership functions |

| VL | VH | ME | LO | VL | HI | LO | LO | VL | Corresponding membership functions |

Figure 4.9. Decoding binary codes into membership functions of rule consequences

The advantage of posing the problem in this way is that the resulting rules are linguistically sound in that both their premises and consequences can bear linguistic meaning. This however compromises its numerical accuracy compared to the least squares training algorithm. This is a price that has to be paid in the transformation of knowledge from a numerical to a linguistic level.

Clustering techniques

Clustering techniques can also be used to train fuzzy rules from training data (see Kosko, 1992). After clusters are identified in the hyperspace of the input vector and output of the training set, each cluster can be projected to the corresponding axes to construct fuzzy sets. These input and output fuzzy sets can then be connected in the form of fuzzy rules. Clusters can be identified using some technique such as self-organizing feature maps (Kohonen, 1990).

4.2.5. The fuzzy alpha-cut technique

The fuzzy alpha-cut technique uses fuzzy set theory to analyse uncertainty in the parameters of a model. Uncertain parameters are considered to be fuzzy numbers with some membership functions. Figure 4.10 shows a parameter P represented as a triangular fuzzy number with a support equal to A_0. The wider the support of the membership functions, the higher the uncertainty. The fuzzy set that contains all elements with a membership of $\alpha \; \varepsilon \; [0,1]$ and above is called the *α-cut* of the membership function. At a resolution level of α, it will have a support of A_α. The higher the value of α, the higher the confidence in the parameter (Li & Vincent, 1995).

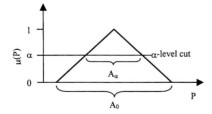

Figure 4.10. Fuzzy number, its support and α-level cut

The method is based on the *extension principle*, which implies that functional relationships can be extended to involve fuzzy arguments and can be used to map the dependent variable as a fuzzy set (Zadeh, 1975). In simple arithmetic operations, this principle can be used analytically. However, in most practical modelling applications, relationships involve partial differential equations and other complex structures that make analytical application of the principle difficult. Therefore, interval arithmetic is used to carryout the analysis by cutting horizontally the membership function at a finite number of α-levels between 0 and 1. For each α-level of the parameter, the model is run to determine the minimum and maximum possible values of the output. This information is directly used to construct the corresponding fuzziness (membership function) of the output, which is then used as a measure of uncertainty.

If the output is monotonic (increasing or decreasing) with respect to the dependent fuzzy variable(s), the process is rather simple since only two simulations are sufficient for each α-level; one for the minimum and another for the maximum value of the parameter at the α-cut. Otherwise, optimisation routines have to be carried out to determine the minimum and maximum values of the output fuzzy membership function for the range of the input corresponding to each α-level. This is because the output corresponding to the maximum and minimum value of the α-level of the input may not necessarily correspond to the maximum and minimum value of the α-level of the output fuzzy set. The fuzzy α-cut technique has been applied by Schulz & Huwe (1997, 1999) to model soil water pressure in the unsaturated zone subject to imprecise boundary conditions and hydraulic properties.

Referring to the principle of uncertainty invariance discussed in the preceding chapter, the fuzzy α-cut technique can be compared with the most widely used Monte Carlo simulation technique (see Chapter 2). In Monte Carlo simulation, an input parameter P subject to uncertainty is considered as a random variable $\mathbf{P}$. A number of realizations P_i of $\mathbf{P}$ is generated and the deterministic model is run for each of them, hence producing an output R_i. The set of outputs R_i represents the set of realizations of the random variable $\mathbf{R}$. The statistical properties of $\mathbf{R}$ are therefore computed from the realizations R_i.

4.3.   Case study: Fuzzy alpha-cut versus Monte Carlo simulation in assessing parameter uncertainty

Parameters of physically based models bear some meaning and can be determined using *in-situ* measurements, calibration, expert judgment, etc. However, the values of these parameters may be subject to significant uncertainty, due to the lack of measurement points, over-calibration or imprecise expert judgment. Uncertainty in model parameters is one of the main causes of uncertainty in model outputs. Uncertainty arising from model parameters can be analysed using several techniques. In this case study Monte Carlo simulation and fuzzy α-cut technique are considered. The Monte Carlo simulation technique treats any uncertain parameter as random variable that obeys a given probabilistic distribution. Consequently, the model output is also a random variable. This technique is widely used for analysing probabilistic uncertainty. The fuzzy α-cut technique treats uncertain model parameters as fuzzy numbers that can be manipulated by fuzzy operators.

This case study aims to compare the features, advantages and drawbacks of the two techniques when applied to the analysis of uncertainty in ground water contaminant transport modelling. The decay rate of the contaminant is considered to be the uncertain parameter. It is also intended to compare the results of the two analysis techniques in relation to the principle of uncertainty invariance. In order to provide a fair basis for comparison, the shape of the membership function used in the fuzzy α-cut technique is the same as the shape of the probability density function used in the Monte Carlo simulation.

The case study is adapted from Abebe *et al.* (2000a).

4.3.1. Groundwater solute transport model

In the presence of advection, hydrodynamic dispersion and first-order kinetics degradation in the liquid phase only, the two-dimensional equation that describes contaminant transport in an aquifer is:

$$\frac{\partial}{\partial t}\left(h\theta C\right)+\nabla \mathbf{F} = RC_s - \lambda h\theta C$$

(4.20)

where $\mathbf{F} = h\mathbf{v}C + h\delta\mathbf{v}\nabla C$

and where C (kg/m^3) is the contaminant concentration in the aquifer, C_s (kg/m^3) is the contaminant concentration of the recharge water, h (m) is the aquifer thickness, $\mathbf{v}$ (m/s) is the Darcy velocity vector, R (m/s) is the recharge rate, δ (m) is the dispersivity tensor, λ (s^{-1}) is the degradation rate, θ (dimensionless) is the aquifer porosity and ∇ represents the divergence operator when applied to a vector and the gradient when applied to a scalar. Usually, the typical order of magnitude for the dispersivity in non-fractured aquifers is 10 m or smaller (Gelhar *et al.*, 1992). Since the distances D (m) dealt with in this case study are one or several kilometres, the Peclet number Pe = D/δ is much greater than unity, which indicates that the predominant phenomenon in this type of problem is advection. The dispersive terms in equation (4.20) were therefore neglected.

Equation (4.20) was solved numerically using a finite volume, Godunov-type method (Guinot, 2001), adapted for the two-dimensional computation of passive contaminant transport in two dimensions.

4.3.2. Study area and data

The two approaches described above were applied to contaminant transport modelling in groundwater for the Vannetin basin in France. This catchment is a sub-catchment of the Grand Morin basin, which contributes to the water supply of the urban area of Paris (Figure 4.11). Water is mostly taken from the main rivers that flow east of Paris. Since these rivers drain a wide aquifer system, the quality of drinking water depends directly on that of water in the aquifer system.

This basin is used extensively for agricultural practices. Regular application of pesticides over the past 35 years has lead to increased pesticide concentration in the aquifers and consequently in the river systems. Therefore, modelling contaminants in the groundwater is important for the management of the surface water quality.

The behaviour of the aquifer flow in the neighbourhood of the aquifer wells in the town of Choisy-en-Brie has already been investigated in a study by Guinot (1995). This investigation was carried out with the aim of defining protection perimeters for the groundwater wells against contamination by pesticides. A physically based model of the Vannetin catchment was built using the MIKE SHE (Abbott *et al.*, 1986a,b) modelling system. The model predicted a quasi-steady flow regime in the Champigny aquifer, where the pumping wells are located. This is due to the fact that the Champigny aquifer is separated from the ground surface by a top aquifer and by a semi-permeable marn layer. Figure 4.12 shows the computed water table.

Grand Morin

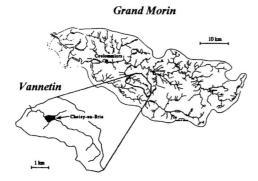

Figure 4.11. Geography of the test site

The study consisted of analysing the effect of a point-wise contamination of the aquifer by Atrazine. Atrazine is a pesticide used for protecting maize. The injection point is indicated by a square on the figure.

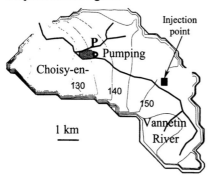

Figure 4.12. Computed aquifer water table. Heads are given in metres

Although the degradation rates of most pesticides in the unsaturated zone are quite well known, very little is known about their behaviour in the saturated zone. Their half-life is generally estimated to be two or three years. On the basis of previous studies carried out on uncertainty assessment in the unsaturated zone (Carsel *et al.*, 1988), the assumption was made of a triangular probability density function for the degradation rate. The mean value for the degradation rate was taken equal to $2.2 \times 10^{-8} \text{s}^{-1}$, which corresponds to a half-life of 2 years. The lower bound of the probability density function is $1.64 \times 10^{-8} \text{s}^{-1}$ and the upper bound is $2.7 \times 10^{-8} \text{s}^{-1}$ (Figure 4.13).

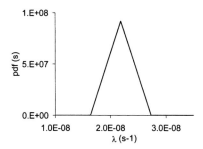

Figure 4.13. Probability density function for the degradation rate

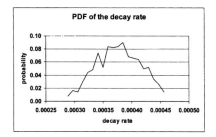

Figure 4.14. Histogram of the degradation rate derived from the sample obtained with a random number generator

From this theoretical distribution, 500 values were generated using a random number generator Obviously more samples could be generated in order to get a smoother distribution. This, however, it is intended to highlight the difference between the two techniques. The histogram of the generated sample is shown on Figure 4.14. Each of these 500 values was taken as input for the transport model described by equation (4.20). In these simulations, the correlation distance of the degradation rate was assumed to be infinite. This means that the value of λ was assumed to be the same everywhere. This assumption was made for the purpose of unbiased comparison with the fuzzy approach that assumes the fuzzy variable (i.e. its membership function) to be the same at every point in space. As shown by previous studies, the assumption of infinite correlation leads to an overestimation of the effects of uncertainty (Guinot, 1995). The membership function of the degradation rate that was used for the fuzzy technique is shown on Figure 4.15.

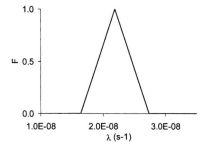

Figure 4.15. Membership function of the degradation rate

4.3.3.   Results

The analyses were carried out in two ways: (1) spatial analysis where the uncertainty over whole model domain is compared, and (2) point-wise analysis where the density probability function (for the Monte Carlo simulation) and the membership function (for the fuzzy α-cut technique) of the concentration are analysed at a given point.

Spatial analysis

In order to evaluate the spatial distribution of uncertainty, it is necessary to establish a measure of uncertainty for the two methods. For the Monte Carlo simulation, such a measure is considered here as the ratio of the standard deviation to the mean concentration of the solute at each grid cell. The measure of uncertainty used for the fuzzy α-cut technique is the ratio of the 0.1-level support to the value of the concentration for which the membership function is equal to 1, *d/d'*, (see Figure 4.16).

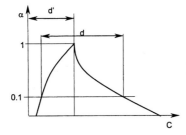

Figure 4.16. Measure of uncertainty for the fuzzy α-cut technique (the ratio d/d')

Figure 4.17 shows the measure of uncertainty obtained by the Monte Carlo simulation, whereas Figure 4.18 shows the results obtained with the fuzzy α-cut technique. The results indicate that the relative width of the fuzzy membership function of the output and the ratio of the standard deviation to the mean of the concentration is an increasing function of the distance to the injection point. The magnitudes of the measures of uncertainty in both cases are different but they convey similar information about the distribution of uncertainty.

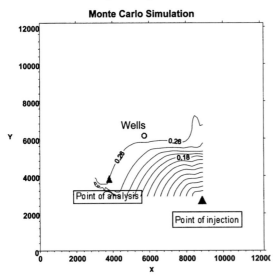

Figure 4.17. Measure of uncertainty for the Monte Carlo simulation (distances in m)

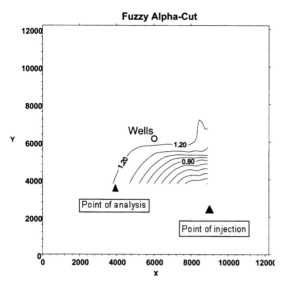

Figure 4.18. Measure of uncertainty for the fuzzy α-cut technique (distances in m)

Point-wise analysis

Figure 4.19 shows the uncertainty at the selected point of analysis. The standardised frequency distribution of the concentration obtained from Monte Carlo simulation is plotted on the same set of axes as the fuzzy number representing the concentration obtained from the fuzzy α-cut technique. Figure 4.20 shows the distribution function and the standardised-integrated fuzzy number.

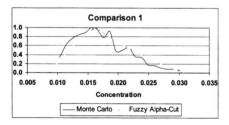

Figure 4.19. Standardised PDF and fuzzy membership function of the output at the analysis point

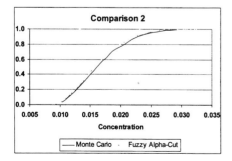

Figure 4.20. Cumulative distribution of the PDF and standardised and integrated membership function of the output at the analysis point

The analysis at the selected grid point indicates that a symmetrical uncertainty in the decay rate results in an asymmetrical uncertainty in the solute concentration. This is a reflection of the effect of the equations used by the model. The width of the output membership function (fuzzy number) is the indication of the sensitivity of the model to this parameter.

Both methods have shown comparable results when the integrated membership function and the cumulative frequency distribution are used for comparison. However, when probabilistic density functions are compared to membership functions, there is a clear indication of the variability in the case of Monte Carlo simulation and consistency in fuzzy α-cut technique.

4.3.4.   Conclusions and discussion

The Monte Carlo simulation approach needed 500 model runs corresponding to the number of samples. This number could be reduced, but only at the expense of the smoothness of the generated frequency distribution and the accuracy of the resulting probability density function. The fuzzy α-cut technique needed only 20 model runs. This indicates an obvious advantage of the fuzzy α-cut technique over the Monte Carlo simulation for this particular case study. However, it should be noted that the solution of the advection-degradation equation is monotonic with respect to the degradation rate. If the results were not monotonic with respect to the uncertain parameter (e.g. in the case of diffusion problems), it would be necessary to use an optimisation technique to construct the fuzziness of the output. The approach would then become more time-consuming than in the present case. Such problems have not been investigated in this case study.

The drawback of Monte Carlo simulation for the present application is that it is time consuming, due to the large number of simulations needed to achieve a satisfactory smoothness and accuracy of the results. On the other hand, it is possible to generate random fields for the uncertain parameter that take into account the spatial structure of these parameters (e.g. shape of the variogram or correlation distance). The fuzzy α-cut technique presents a strong alternative to the Monte Carlo approach. It is faster for applications where the output is a monotonic function of the uncertain parameters, but the effect of non-monotonicity of outputs with respect to parameters has to be investigated in terms of computational effort. Its drawback is that, so far, it is applicable only under the assumption of infinite correlation distances.

Generally, both methods gave similar results provided that the correlation distance of the decay rate is assumed to be infinite, thus conforming to the principle of uncertainty invariance. However, particular details of analysis, computational time and representation of uncertainty are different, which may lead to the choice of one method or another depending on the nature of the problem.

CHAPTER 5. COMPLEMENTARY MODELLING

This chapter is intended to instigate the methodological components of the thesis, which are information theory and artificial intelligence, to address uncertainty in physically based models. The art of integrating the techniques to the problem is termed complementary modelling. As its name implies, the methodology is based on a complementary use of physically based models that are based on knowledge of the constituent physical processes and intelligent data-driven models that are capable of learning directly from data. The applicability of the methodology and the validity of the arguments that led to it are then tested with hypothetical and real-life problems of varying complexity in the later parts of the chapter.

5.1. Model updating practices

When a physically based model is used for forecasting purposes, recently observed data can be used both to refine the model and its results in various ways. For stream flow prediction models, O'Connell & Clarke (1981) and Refsgaard (1997) identified four types of updating procedures. These correspond to the input data, state variables, parameters, or the output of the model. The choice of the updating mechanism generally depends on what is considered to be the main cause of the discrepancy between observations and model predictions.

Input updating

Input updating refers to the alteration of the input data of a model from what is actually observed. It is a common practice, for example, to average the precipitation data from gauges or to accumulate precipitation data over a time period before feeding it into a catchment model. From the point of view of model updating, however, input updating refers to the alteration of the input data of a model in such a way that model outputs fit the corresponding latest observations. The argument behind input updating is that the input data might be the main source of uncertainty.

Parameter updating

Another way to use the latest observations in improving the accuracy of model forecasts is by updating the parameters of the model. This generally follows a procedure that links the discrepancy between model forecasts and the corresponding latest observations to all or some of the parameters of the model. Babovic *et al.* (2001), commented that parameter updating can be considered as a re-calibration problem repeated as new measurements come in. Yang & Michel (2000) used parameter updating to forecast flooding with a conceptual watershed model. Their approach updates the parameters of the model around a baseline set of parameters with the intention of reducing the error in the latest stream flow forecasts as close to zero as possible.

Parameter updating can be tiresome especially in real-time forecast models. The procedure could mean altering a representative set of parameters calibrated over a long historical record because of a mismatch in the latest observations. Another drawback with parameter updating is that all the errors are attributed to parameter uncertainty. This assumption ignores the possible presence of inherent deficiencies, say, in the structure of the model.

State updating

Recently observed data can be used to improve the state variables of a model to subsequently obtain better model forecasts. This is generally performed by relating the temporal variation of a state variable of the model to an external observation. If an observation is done at a time t and its value does not exactly fit with the model state variable, the state variable can be forced to a new value in order to bring the model prediction closer to the corresponding measured value. The model is then allowed to evolve freely, from this new initial state, until another observation is available.

The most widely used method of state updating is Kalman filtering (Kalman, 1960). The Kalman filter involves calculating a correction term taking into account the differences between the model predictions and observations, and locally linearising the model. With more accurate observations, the *a posteriori* state variables can be more accurately defined. The corrections on different state variables, using different observations, can also be linked together. The Kalman filter is based on the state space-time domain formulation of the process involved. It takes into account the system dynamics, maintaining the underlying stochastic behaviour. One of its advantages lies in the fact that it helps to update outputs at locations where little or no measurements are available. The complexity of Kalman filtering generally depends on the complexity of the model since it directly targets the state variables.

Lee & Singh (1998) applied Kalman filtering to update the state variables of the Nash model in order to obtain an improved runoff prediction. Its application to hydrodynamic modelling has been explored by Verlaan (1998) and Canizares (1999). A variety of applications of Kalman filtering for water related models is discussed in Chiu (1978).

Updating model outputs

Updating the outputs of a model can be done directly or indirectly. In this context, direct updating refers to the use of a separate model to improve the initial outputs directly on the basis of the latest observations. Indirect output updating refers to the prediction of the expected errors in the initial outputs in order to subsequently update them. Indirect updating is also called *error prediction*. An obvious advantage of updating model outputs is that it bypasses the main process model. Unlike updating inputs, parameters or state variables, updating the model outputs does not involve any additional runs of the primary model. This is particularly important in applications involving a combination of computer-time demanding model and real-time forecasting. The other advantage is that it can be applied regardless of the nature of the primary model. Refsgaard (1997) draws attention to the importance of updating forecasts by error modelling rather than updating the parameters or the state variables in cases where the source of model errors is unknown.

Output updating is generally performed with a separate data-driven model. Neural network models have been a favourite choice in recent literature. Babovic *et al.* (2000) performed a comparative study of different techniques to predict errors from a 2-D hydrodynamic model of the Adriatic Sea and the Venice lagoon. Abebe & Price (2000) showed that errors arising from an inappropriate selection of computational time steps could be reproduced using ANN models. Babovic *et al.* (2001) used ANN models to update water levels and currents forecasted by a 2-D model of the Oresund Strait between Denmark and Sweden. Shamseldin & O'Connor (2001) used ANN models to update runoff forecasts by using the simulated flows from a model and the current and previously observed flows as input, and the corresponding observed flow as the target output. The approach was applied to update daily flow forecasts for a lead-time of up to four days. They reported that ANN models gave more accurate improvements than autoregressive models. Also Lekkas *et al.* (2001) showed that error forecasting provides improved real-time flow forecasting, particularly when the forecast model is poor. More studies conducted by Abebe & Price 2002(a, b) have shown the importance of complementary modelling in rainfall-runoff and flood routing models in improving the accuracy of forecasts.

5.2. The complementary modelling concept

Literally, complementary modelling refers to the process of complementing one model by another to obtain a more 'complete' model. The generation of a complete model is practically unrealistic. The most realistic expectation is a better or more representative model. In the context of this thesis, complementary modelling is presented as a process of bridging the gap between a physically based process model and the actual physical process by the use of a separate data-driven model. The model developed to represent the physical process is referred to as the *primary model* whereas the model intended to bridge the gap between the primary model and the physical process is referred to as the *complementary model*.

A straightforward way to apply complementary modelling is to predict the errors of the primary model using a separate model and subsequently updating the predictions. In this sense the complementary model *complements* the physically based model (see Figure 5.1). A successful prediction of the expected errors of a physically based model improves forecast accuracy and consequently reduces the associated uncertainty since residuals can be considered as indicators of the discrepancy between a physically based model and the physical process it intends to represent.

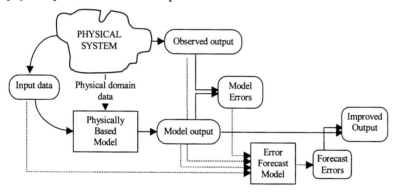

Figure 5.1. A complementary modelling approach (adapted from Abebe & Price, 2004)

A problem associated with error prediction as a sole means of managing uncertainty is that there is no guarantee that the residual errors of a physically based model are predictable by a complementary model. Patterns in errors are not always systematically and deterministically related to other parameters. In addition, direct prediction of the expected errors by a separate model might not be the way the modeller intends to manage the uncertainty.

However, prediction of the expected errors of the primary model is not the only way in which the primary model can be complemented by a separate model. Provision of additional information regarding the accuracy of predicted outputs is also complementary modelling. In fact, the complementary model can be used to characterize the expected accuracy of the primary model. This can, for example, be a linguistic description of the prediction accuracy, the expected bias of the errors, or confidence intervals in which the predictions fall with a predefined likelihood, all pertinent to the state under which the model prediction is generated. In each case the complementary model needs data that are relevant to predict or characterize the expected accuracy of the model. In this respect, error modelling describes the function of the complementary model more generally than does error prediction.

5.3.   Model uncertainty as an error modelling problem

A question that can arise here is how model uncertainty can be treated as an error modelling problem. There are several reasons to justify the use of error modelling as a means of handling model uncertainty.

❑   Historical residual errors between model outputs and the corresponding observed data are the best available quantitative indicators of the discrepancy between a model and the world it is intended to represent. The greater this discrepancy is the greater the uncertainty associated with the model. A consistent and systematic reduction of the discrepancy results in a subsequent reduction in the uncertainty associated with the model and its results. The contribution of errors in data to the overall uncertainty affects the strict use of residual errors as a sole measure of the gap between the process and its model. However, for practical purposes, data uncertainty is also a part of the total uncertainty in modelling and there is a good chance that a complementary model can represent its systematic component.

❑   In general, the smaller the residual a model generates over the test data, the more confident the modeller becomes about the closeness of the model in reproducing the actual physical process. Once the basic physical principles and theoretical background is embedded in a model, one of the intentions of validating a model is the close reproduction of the observed equivalent of the model output as closely as possible. Absolute reproduction of observed data is almost impossible for models of natural systems and thus errors hardly ever disappear.

❑   The complementary modelling approach treats uncertainty as modelling the total discrepancy between the model and the process no matter what the actual causes of the discrepancy are. Therefore, it helps to minimize the overall uncertainty compared to most of the traditional techniques of managing uncertainty, which generally address specific aspects of uncertainty in modelling caused by parameters, model structure, input data, etc.

The other question that can arise in this respect is how successful can error modelling be? The success of error modelling depends of the presence of recoverable patterns in the historical error time series of a model, which in turn depends on factors such as:

❑   How good the prediction model (its structure and parameters) represents the processes involved, and

❑   The quality of the boundary and static data used by the model (accuracy, resolution, noise/signal ratio, etc) as well as the target data, which the model intends to predict, since it is used in the computation of the errors.

The process of detecting these patterns is more of an art than an exact science. The success of detecting any recoverable pattern in the error time series depends, among other things, on:

❑   The availability of data for such analysis,

❑   The proper use of physical insight in trying to relate the state variables to the error time series,

❑   The robustness of the analysis techniques applied, and

❑   The exhaustiveness of the search for these patterns,

❑   Good-old-fashioned luck!

5.4.   Information theory in complementary modelling

When an error time series contains a visually observable pattern such as errors being in phase or having a similar amplitude ratio, the modeller can track them back to their probable cause relatively easily, and can even take steps to improve the physically based model accordingly. Serban & Askew (1991) sited three typical forms of error patterns observed in hydrographs: phase errors, amplitude errors, and shape errors (Figure 5.2). These are patterns that can be visually recognized and only in the time domain. What if a combination of these patterns appears? What about error patterns that do not fall in any of these categories yet have a systematic presence in a domain other than time, such as in space or in some of the boundary data or model state variables? Under such circumstances, visual identification of their probable cause becomes impossible and the chances are that the errors will most likely be non-linear with respect to time and the state variables of the model. Therefore, it becomes as important to use analysis techniques that are capable of detecting non-linear as well as linear relationships, as it is important to use modelling techniques that can handle non-linear as well as linear relationships between data.

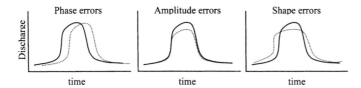

Figure 5.2. Types of errors (based on Serban & Askew, 1991)

It is here that the measure of the degree of relationship between data sets plays an important role. A number of methods can be used to analyse the relationship between two data sets. The two methods that are frequently used in this thesis are cross correlation and average mutual information (AMI) which are described in Chapter 3.

The coefficient of correlation can be used as a measure of the relationship between two data series by applying some type of regression function. The function might be linear, polynomial, exponential, etc. Linear correlation is the one most widely used. When two data series can be absolutely correlated by this function, the correlation coefficient becomes unity. The advantage of correlation is that it has a known absolute maximum and minimum. Its disadvantage is that the functional relationship between the two data series is unknown *a priori*. In most cases, the relationship between two data series might have a linear component; nevertheless, it is the presence of non-linearity that makes the problem more challenging. It is important to note that when dealing with the relationship between the residual errors of a prediction model and other data series, the expected relationship is seldom linear. This is mainly because a linear bias in residual errors can be detected and accounted for at the model calibration stage, relatively easily.

The information theory-based measure AMI is free from the main drawback that cross correlation has in that it does not depend on any predefined function. This is because it is based on set theory and is essentially computed using the individual probabilities and the joint probabilities between data series. The fact that in practice data are monitored at discrete intervals rather than on a continuous basis urges the use of class intervals to compute these probabilities. Since the choice of class intervals (or the number classes) affects the resulting probability distributions, it subsequently affects the AMI. It can be proven that when the number of class intervals is increased, the degree of freedom increases and so does the entropy and the AMI. Unlike the entropy, the AMI converges with increasing number of class intervals (see Kantz & Schreiber, 1997). However, limited amount of observed data often dictate the class interval to be set at a practicable limit. This perhaps is the only drawback of the AMI measure.

Due to its capability in identifying non-linear relationships between data, the AMI measure is extensively applied in this thesis to detect how much residual information is available in historical model prediction errors. It helps to select the data that are best related to the errors and can be used in setting up an error prediction model. In addition, it helps to identify the data that are redundant in the case that simplified models are needed.

5.5.   Data-driven models in complementary modelling

Thanks to recent advances in computing capacity, a variety of data-driven modelling techniques with intelligent learning algorithms have nowadays been developed and applied successfully to various problems. The combined application of physically based and data-driven models is not new. For instance, the coupled use of conceptual and data-driven models is discussed in Young (2001) as yet a different class of models named Hybrid Metric-Conceptual models. However, the way in which the two classes of models complement each other in the problem can be different.

In this study, the complementary model is developed using data-driven modelling techniques. One of the reasons for this is the fact that physically based models are based on physical principles whereas data-driven models exploit the relationship between data, which gives the two classes of models a complementary nature.

The second reason is that, considering the discrepancy between the primary model and the actual physical process as a separate process, model errors can be considered as data observed from this 'process'. It is for this process that a complementary model is intended. This residual process is difficult to handle with a physically based model since the primary model has used most of the tractable physics. The next remaining option would be to consider a modelling technique that exploits the data rather than the physical insight, which is a data-driven technique.

5.6.   The need for complementary modelling

A possible ethical concern regarding complementary modelling is that it may divert the attention of the modeller from the effort of refining the physically based model. However, with the available physical knowledge about a process, there is a limit beyond which a physically based model cannot be improved. Sometimes, the improvement might not justify the cost needed to improve the primary model or collect more physical domain data compared to the effort needed to develop a complementary data-driven model. Under such circumstances, complementing a physically based model with a data-driven model becomes useful. After making full use of the available physical knowledge, the modeller should turn to modelling the residual process, which is described by the mismatch between the historical data and the corresponding model predictions, and a modelling technique that can best utilize the resulting data, which is data-driven modelling.

Some of the advantages of complementary modelling are as follows:

❑   The complementary modelling approach, as it is formulated in this thesis, follows an external and parallel integration of the two models, meaning that the primary model is independent of the complementary model. Because of this, the procedure does not involve any additional runs of the primary model other than what would be necessary without complementary modelling. In fact, it can be applied in such a way that it does not affect the way the primary model is routinely operated. This is particularly important in the case of existing operational models that would need considerable investment to update.

❏ For very complicated physically based models with spatially numerous outputs, only selected outputs of interest can be updated using a complementary model, which gives complementary modelling a competitive edge in operational real-time forecasting models compared to updating techniques that involve all outputs of a model. For example, a coastal hydrodynamic model computes water levels at all grid points. A complementary model can be trained to update water levels only at selected points.

❏ Any relevant data can be integrated indirectly into initial forecasts produced by a physically based model. Sometimes, a physically based model may not use important data only because it does not have provisions to use these data. By the use of complementary modelling, it would be possible to make use of relevant data that are not part of the primary model.

❏ The primary model, which in this context is a physically based model, is based on physical principles represented by a mathematical formulation. A separate model with a learning capability could indirectly handle the uncertainty resulting from missing or ignored processes in the primary model.

❏ In addition to their growing development and applicability, data-driven modelling techniques with learning capabilities such as autoregressive models, transfer functions, local linear models, neural network models, and fuzzy rule-based models, are increasingly available both in commercial and research domain.

5.7. Model calibration versus complementary modelling

The task of developing models sometimes involves a stage of calibration. Calibration is the process of adjusting the parameters of a model, such as roughness coefficients for a river model, so that the model reproduces observed data to an acceptable accuracy. Calibration is a logically acceptable step in model development, particularly when it is done to determine the values of parameters, which for some reason cannot be measured from the physical system. However, there is some controversy surrounding calibration, particularly on how and to what extent it should be done.

If model structure is defective and the calibration data is not 'representative', the calibration process can force the parameter values to go beyond their true values. The word 'true' is used here on the assumption that even parameters that cannot be measured have values that correspond to their physical meaning. It has been stated in Cunge (2003) that calibration must be limited to parameters that are *invariant* between the development (instantiation) and application (exploitation) stages, unless the purpose is to study the sensitivity of the model to modifications in its parameters. However, in practice, calibration is sometimes applied as a means of obtaining close reproduction of observed events. In the context of river models, *Cunge et al.* (1980) stated that close reproduction of recorded hydrographs is not a sufficient proof of the reliability of a model; another model may be more reliable even though its reproduction of observed events is not quite as good. Judging whether parameters values are correct is not always possible because, particularly in conceptual models, which are not fully physically based, even the range of representative values of parameters might not be known.

The contribution of incorrect parameter values to model prediction errors is most likely systematic, implying that there is a good chance of developing a complementary data-driven model that can predict these errors. This is perhaps the reason for the argument of Lekkas *et al.* (2001) that error forecasting provides improved real-time flow forecasting, particularly when the forecast model is poor. A model with poor structure and incorrect parameters is likely to generate more systematic errors than one with more representative structure and parameters.

The prospective use of complementary modelling raises the question as to whether a model can be inadequately calibrated because of relying on the complementary model for the potential consequences. The position in this thesis regarding model calibration is that the physically based model has to be developed in such a way that its parameters assume physically acceptable values. If calibration is needed to achieve this, then it should not be avoided because there is going to be a complementary data-driven model. However, calibration has to be limited to the purpose of fixing the parameters rather than obtaining an exact reproduction of observed values at the expense of the physical meaning of the parameters. If the primary model has to remain 'physically based', its structure and parameters have to reflect the physics of the system as closely as possible. From this point of view, it is better to have a representative model and live with the appreciation of the uncertainties it might have than having a model with incorrect parameters but that reproduces observed values more closely.

5.8.   Formalizing the complementary modelling procedure

5.8.1.   Defining model errors

Model prediction errors, or errors, as referred to in parts of this thesis, lie at the centre of the complementary modelling approach. Therefore, it is necessary to define errors formally.

In the context of this thesis, an error is a measure of the mismatch between a predicted value and the corresponding observed or 'true' value. A model M is intended to predict y using a number of inputs $x_1, x_2, ..., x_n$ can be defined as:

$$y' = M(x_1, x_2, ..., x_n) \tag{5.1}$$

where y' is the model predicted estimate of y.

The model error E is then a function of y and y' intended to shown the mismatch between them. The simplest and most widely used definition of the error is the difference between y and y' which can be written as:

$$E = y' - y \tag{5.2}$$

Defining errors in the form of the difference takes into account the algebraic sign of the error. Sometimes, the algebraic sign is not important and the error can be defined as:

$$E = |y' - y| \tag{5.3}$$

In some cases, it might be needed to define the error in a dimensionless form relative to the observed value as:

$$E = |y' - y| / |y| \tag{5.4}$$

This definition of errors does not have to be confused with the definition used in measurement errors where absolute error refers to the difference whereas the dimensionless form is known as relative error.

In all of the above definitions, the mismatch between y' and y is contained in the expression $y' - y$. However, the mismatch can as well be defined as a ratio such as:

$$E = y' / y \tag{5.5}$$

Care has to be taken in defining errors as a ratio since the error can be undefined when y is zero. In that case, it may be needed to standardize the denominator of the expression.

In general, for the purpose of complementary modelling, the error can be defined in any form. However, it has to satisfy the following two conditions:

(1) The error term must reflect the mismatch between the model predicted value and the observed value, and

(2) It must be possible to rewrite the error function in such a way that y can be defined as a function of the error and the predicted value y' so that predicted errors can be used to update preliminary outputs. In other words, there must be a function ζ that can be defined as:

$$y'' = \zeta(E', y') \tag{5.6}$$

where E' is the expected error and y'' is a better approximation of y than is y'.

5.8.2.   Data for complementary modelling

A combination of historical data and the corresponding historical predictions of the primary model can be used to develop and validate the complementary model. The actual data needed for this purpose can vary depending, among other things, on the type of the model and on data availability. A combination of physical insight and data relationship analysis may be used to select relevant data for the complementary model. The input data that can be considered for use in developing the complementary model include:

❑ Input data to the primary model

❑ Selected state variables

❑ Intermediate or final outputs of the primary model

❑ Target data of the primary model

❑ Historical prediction errors of the primary model

❑ Other data from the physical system that, according to physical insight or data relationship analysis, are found to be important in reproducing the errors of the primary model.

❑ Time and space derivatives of input data and state variables of the primary model

5.8.3.   Data relationship analysis

After gathering the necessary historical data about the process under consideration and the corresponding performance of the primary model, analysis of the relationships between data is the next logical step in applying the complementary modelling procedure. Such an analysis helps to detect the presence of recoverable patterns in the historical residuals of the model. Appropriate analysis techniques include AMI, cross correlation, Fourier transformation, etc. combined with physical insight. They can be used to detect the degree of relationship between the residual errors of a physically based model and the relevant data. A schematic illustration of the data relationship analysis is given in Figure 5.3. These techniques can help to check periodicity, time dynamics and degree of relationship between residual errors and other parameters.

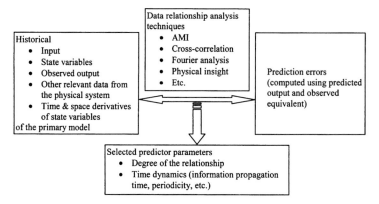

Figure 5.3. Data relationship analysis

5.8.4.   Developing the complementary model

The outcome of the data relationship analysis is often a list of selected parameters that are best related to the residual errors of the physically based model. It is also possible to know the degree to which they are related based on AMI measures. This information can be used to develop the complementary data-driven model. The process of developing the complementary model is essentially the same as the process that the selected data-driven modelling technique dictates.

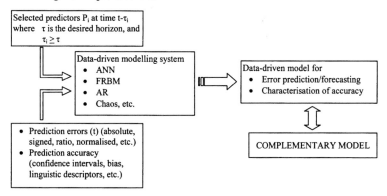

Figure 5.4. Complementary model development

The procedure of developing the complementary model is depicted in Figure 5.4. As shown in the figure, the procedure has two inputs:

One of the inputs is the historical prediction accuracy, which can be in the form of prediction errors or in some other form such as bias, confidence intervals, linguistic description of accuracy, etc. In either case, it is defined as a function that reflects the mismatch between model predictions and the corresponding observations.

The other input is a list of selected predictors P_i that are found to be best related to the prediction accuracy. In forecasting applications, care has to be taken to make sure that the selection of predictors is carried out maintaining the desired forecast horizon, τ, in which case $\tau_i \geq \tau$ must be satisfied, where τ_i is the lead-time (separation time) between the prediction accuracy and the predictive parameter P_i.

These two inputs are fed into the data-driven modelling system, the outcome of which is a data-driven model that is capable of predicting the expected accuracy of the primary model. The resulting model is essentially the complementary model.

A complementary model for the purpose of error prediction can be developed using the following procedure:

(1) Prepare the historical errors, inputs and output data series of the physically based model. Also prepare derivatives and external data that, based on physical insight, are related to the residual errors.

(2) Evaluate the information content within the residual errors and with other data selected in step (1). In cases involving a time series, the information content can be analysed by varying the lag times between the data involved.

(3) Select the predictive parameters that are best related to the errors based on the information content. Also select the lag times at which the information content is a maximum. If the primary model is intended for forecasting purposes, care has to be taken to ensure that the input data to the complementary model are available in forecast mode.

(4) Develop a data-driven model that maps the selected predictive parameters to the residuals. This step depends on the type of the data-driven modelling approach selected for the complementary model.

The procedure of applying complementary modelling in operational settings is straightforward. At first, a forecast is made using the primary model in the usual way followed by the complementary data-driven model. If the complementary model is intended for error forecasting, then the primary model forecasts are subsequently updated. In the case where the complementary model is intended to forecast the accuracy of the primary model predictions, say, in the form of bias and confidence bounds or linguistic description, the outputs of the primary and complementary models are presented together, the former as a forecasted value and the latter as an estimate of the expected accuracy of the forecasts.

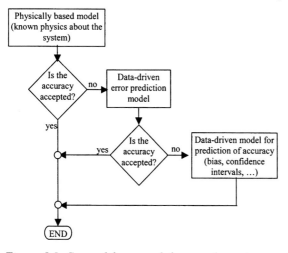

Figure 5.5. General framework for complementary modelling

The general framework of application of the complementary modelling approach is shown in Figure 5.5. When the accuracy of the primary model is not satisfactory, then the complementary model is developed as a data-driven model that directly predicts the errors of the primary model. However, the accuracy of the error prediction model may not be adequate particularly if the errors time series does not contain systematic patterns. In that case, the complementary model can be developed with the intention to predict the accuracy of the primary model such as bias and confidence intervals or linguistic descriptions.

In the following section (§5.9), the applicability of the complementary modelling approach is demonstrated using an artificial neural network to forecast errors of water level of predictions made by a 1-D hydrodynamic model of a hypothetical estuary. Then in §5.10, fuzzy rule-based models are applied to forecast a linguistic description of the accuracy of water level predictions. The methodology is further demonstrated using real life problems of flood routing and rainfall-runoff modelling in §5.11 and §5.12 respectively. Practical applications in river flow forecasting and coastal hydrodynamics are presented in Chapters 6 and 7.

5.9.   Case study: Neural networks to complement hydrodynamic models

This case study is intended to demonstrate the use of neural networks in complementary modelling that involves 1-D hydrodynamic model of a hypothetical estuary. The model has upstream flow and downstream tidal boundary conditions. The case study is adapted from Abebe & Price (2000).

5.9.1.   Introduction

Representing a river system with a mathematical model requires the discretisation of particular partial differential equations to a level which the degree of accuracy of predictions dictate. Generally, low spatial and temporal resolution result in a higher discrepancy from the true solution of the governing equations. In hydrodynamic models, spatial and temporal discretisation go along with rules that arise from the stability criteria of the solution scheme. Implicit schemes generally allow a wider range of computational time steps without losing computational stability. However, this does not imply that all time steps that result in a stable computation are equally accurate in representing the physical system. In most practical cases, the celerity changes from time to time and grid point to grid point depending on temporal fluctuations in boundary conditions or variations in the cross-section along the reach. This leaves the choice of the best computational time step to the modeller. The best computational time step is often governed by the celerity at which information propagates from one boundary to the other. Obviously, if the spatial grid is fixed, which is usually the case, higher celerity conditions are better dealt by smaller time steps. This criterion often leads to time steps too small compared to the physical phenomena under consideration (Cunge *et al.*, 1980).

Most hydrodynamic modelling systems use a constant time step. However, some advanced systems reduce the time step based on convergence criteria. For relatively large models, this might increase the computational time significantly since the full computation has to be repeated at all points, sometimes involving a reduction of the time step several times.

It is possible to determine analytically the loss of accuracy arising from improper computational time steps due to phase and amplification errors (Cunge *et al.*, 1980) using phase and amplitude portraits. However, such analysis techniques are limited to single wave components and not for all wave components due to the fact that the high non-linearity of the equations governing the flow makes the analysis complicated. Therefore, it is necessary to carry out a quantitative analysis of the loss of accuracy arising from the use of fixed computational time steps.

This case study investigates the level of accuracy lost by using a fixed computational time step and the possibility of reproducing the lost accuracy with a neural network model. The possibility of replicating a hydrodynamic model with an ANN model has been shown by Dibike *et al.* (1998) and Dibike and Solomatine (1999). Here the ANN model is intended to relate the discrepancy at the current time step to the flow conditions at the previous time step. The trained network will work in parallel with the physically based model.

5.9.2. Problem formulation

The physical system

The study is performed on a water level prediction model of a hypothetical estuary 10 km long and 100 m wide with a rectangular cross-section (Figure 5.6). It has a constant bottom slope of 0.001 and a Chezy roughness coefficient of 40. The reach is schematised into 10 longitudinal grid cells each 1 km long (Figure 5.7).

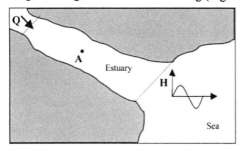

Figure 5.6. The physical system

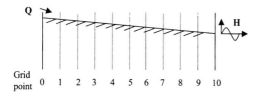

Figure 5.7. Schematisation into grids

The hydrodynamic model

The model is set up using the finite difference solution of the 1-D full dynamic de Saint Venant equations that can be written as shown in equations (5.7) and (5.8).

$$\frac{\partial Q}{\partial x} + b\frac{\partial h}{\partial t} = 0 \tag{5.7}$$

$$\frac{\partial Q}{\partial t} + \frac{\partial}{\partial x}\left(\beta\frac{Q^2}{A}\right) + gA\frac{\partial}{\partial x}(h+H) + gA\frac{|Q|Q}{K^2} = 0 \tag{5.8}$$

where Q = discharge, h = water depth, b = storage width, A = cross-sectional area, K = conveyance, β = Boussinesq coefficient, H = bottom elevation

The equations are solved using the Preissmann (box) scheme (see Abbott, 1979; Cunge *et al.*, 1980). The Preissmann scheme, shown in Figure 5.8, is implicit and numerically stable for all Courant numbers, $Cr = c\Delta t/\Delta x$, where c = celerity, Δt = time step, Δx = space step. The values of ψ and θ determine the location of the centre of the scheme in the grid cell. For $Cr = 1$, the scheme gives the exact solution of the equations. However, phase errors and amplification errors occur for Courant numbers higher and lower than unity, respectively.

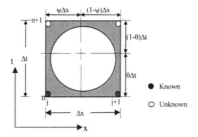

Figure 5.8. The Preissmann solution scheme

Boundary conditions

The estuary model uses the upstream flow and the downstream depth as boundary conditions, respectively. The upstream flow is generated by superimposing the hydrographs resulting from four precipitation events over a base flow. The downstream depth is generated assuming a lunar tide with a period of 12 h and amplitude of 2.5 m at the seaside with a mean depth of 15 m. Both boundary data are plotted in Figure 5.9.

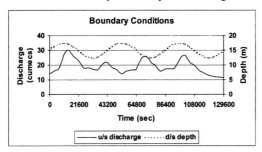

Figure 5.9. Upstream and downstream boundary conditions

Extent of the problem

In order to investigate the extent to which a proper choice of computational time step affects the model, two solution procedures are considered:

1. The first procedure involves the use of a constant time step of 200 s for the whole river reach throughout the simulation period. However it is known that, for values of Courant number different from unity, the solution is subject to amplitude and phase errors.

2. The second procedure involves varying the time step according to the flow conditions, intending to maintain the average Courant number as close to unity as possible. The solution obtained will be close to the analytical solution of the governing equations.

Three cross-sections of the river reach located at grid points 1, 5 and 9 are selected to evaluate the discrepancy in the predicted water level between the two computational procedures, which is also the error as a result of using a constant time step. Figure 5.10, Figure 5.11 and Figure 5.12, and Table 5.1 show the discrepancy in the water level (depth) prediction at the three selected cross-sections.

It can be seen that, even though the discrepancies seem low relative to the total depth, they are quite important compared to the amplitude of the tide. The highest water level prediction error is about 0.90 m, which is rather significant compared to an average depth of 15 m. The next step is to predict these discrepancies using an ANN model.

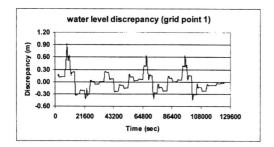

Figure 5.10. Errors in depth computation at grid point 1

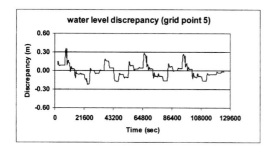

Figure 5.11. Errors in depth computation at grid point 5

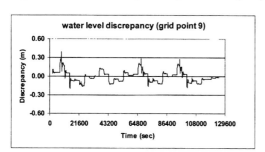

Figure 5.12. Errors in depth computation at grid point 9

Table 5.1. Errors in computation as a result of the choice of time step

Grid point	1	5	9
RMSE (m)	0.2163	0.1031	0.0872
Maximum absolute error (m)	0.9195	0.3619	0.3922

5.9.3. Complementary neural network solution

Input data for ANN

The data for the ANN model is generated in such a way that the depth and discharge at the current time level are used to predict the discrepancy in depth prediction at the next time level. For every time level, simulations are carried out using both procedures mentioned earlier, and the discrepancies in depth prediction are calculated. After sufficient examples are generated, two-thirds of the data is used for training and the rest for verification of the ANN. The pre·lictions obtained in the first 10 time steps are discarded to exclude the effect of the initial conditions. An RBF type neural network is used to represent the relationship between input (depth and discharge at the current time level) and output (depth discrepancy at the next time level). Therefore, the RBF network has 22 input nodes (depth and discharge) and 9 output nodes (errors).

Results

Table 5.2 shows the errors in the prediction of the depth discrepancies using a neural network at the selected grid points. Figure 5.13 shows the scatter diagram of the discrepancies and their prediction using the ANN at grid point 1. In Figure 5.14 the time series of the discrepancies and those predicted by the ANN are presented for the same grid point. This means that the depths computed by the deterministic model with a constant time step of 200 s should be corrected by discrepancies computed by the ANN in order to improve the results.

Table 5.2. RMSE (m) of the prediction of discrepancies using ANN

Grid point	1	5	9
Training	0.0359	0.0227	0.0253
Verification	0.0335	0.0210	0.0241

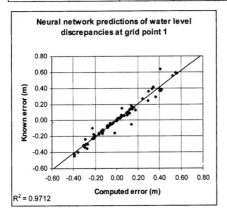

Figure 5.13. Prediction of errors in water level at grid point1 using ANN (verification)

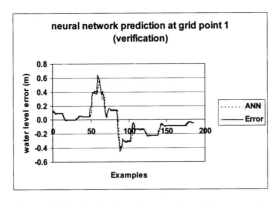

Figure 5.14. Prediction of water level discrepancies shown in Figure 5.10 using ANN

5.9.4.   Conclusions and discussion

The study demonstrated that the use of a fixed time step causes significant errors in the prediction of water levels with a hydrodynamic model. An ANN model of the RBF type can learn the errors caused by the use of a fixed time step. Subsequently, the ANN model can be used to predict errors. This highlights the fact that intelligent models can be used as complementary models to physically based models.

For this test case, errors are obtained by assuming that one of the solution procedures is true. A more reliable option is to test the complementary modelling approach using errors with measured water levels. This will help to investigate whether errors from other causes such as ignored higher order terms and incorrect model parameters can also be accounted for. The ANN model used input data from all the grid points at the previous time step. For large models this will increase the size of the network. A more selective approach has to be used to limit the number of inputs.

5.10.   Case study: Fuzzy rule-based models in complementary modelling

In this section, the possibility of applying a fuzzy rule-based model (FRBM) to characterize the overall prediction uncertainty of a physically based model is investigated. The errors between model predictions and corresponding historical observations are used as the basis for the analysis. The intention is to extract anticipatory fuzzy rules that relate the expected prediction errors to selected prerequisite state variables such as:

"IF *input 1* is LOW AND *input 2* is HIGH THEN *expected error* is POSITIVE HIGH"

where *input 1* and *input 2* are the predictive parameters selected as prerequisite and *expected error* is the consequence.

The FRBM established as a result of such analysis can be applied in real-time forecast systems since the rules can be formulated offline using the historical performance of the model. A series of such rules covering the input domain of a prediction model will help the decision-maker to predict the accuracy even before the corresponding measured values of the target output are available.

An adaptive fuzzy rule-based modelling system along with genetic algorithm is used to extract rules from data. The proposed approach is demonstrated with the water level prediction model of a hypothetical estuary, which is the same model used in the preceding case study. The errors are introduced artificially by altering the computational procedure. The material in this case study is adapted from Abebe & Price (2003a).

5.10.1.   The geno-fuzzy solution procedure

To apply a FRBM to characterize the accuracy of an existing prediction model, there needs to be a set of historical prediction errors of the model. Also needed are the corresponding state variables such as boundary data and/or other relevant data that not only represent the state under which the model predictions were made, but also characterize the prediction errors. The state variables and the prediction errors are used as the input and output of the FRBM respectively.

The approach followed here uses predetermined and linguistically meaningful fuzzy sets on the output as well as the input domain. The training process is then posed as a combinatorial optimisation problem that is intended to map input fuzzy sets to output fuzzy sets. A genetic algorithm is used to perform the optimisation after defining an appropriate payoff function. The payoff function is the inverse of the RMSE between the output of the FRBM and corresponding training data. This training approach is described in §4.2.4. The advantage of posing the problem in this way is that the results are linguistically sound. Training fuzzy rules with predefined membership functions on the output is numerically less accurate than the least squares training algorithm. However, this is the price that has to be paid in transforming knowledge from a numerical to a linguistic level.

The solution procedure has four steps:

Step 1: The state variables (predictive parameters) that are best related to the prediction errors are selected first. This can be done using physical insight and problem specific knowledge. It is also possible to use a cross correlation coefficient. However it is recommended to use the AMI measure since it can help detect non-linear as well as non-linear relationships.

Step 2: Sometimes the selected predictive parameters might be best related to the errors at a delayed time. In this case, it is important to determine the lead-time at which these parameters are best related to the prediction errors. As in the first step, problem specific knowledge can be used. Also the lead-time corresponding to the maximum AMI (or correlation coefficient) between the predictive parameter and the error time series may be used.

Step 3: The predictive parameters and the corresponding prediction errors are used as the input and output respectively of the FRBM. At this stage, the range of input and output data of the FRBM are partitioned into a number of desired fuzzy sets that may be linguistically described, such as LOW, MEDIUM, HIGH, etc. The membership functions corresponding to these fuzzy sets are also defined.

Step 4: With the number of fuzzy sets covering each input space known, the number of rules is also known. For instance, if there are two inputs covered by m and n fuzzy sets, the number of rules is (m x n). Also at this stage the IF part (premise) of all the rules is known. The remaining task is to assign the proper output fuzzy set to the THEN part (consequence) of each rule. Since this is a discrete optimisation problem, genetic algorithm is suitable to find the best output fuzzy set for each rule. The chromosome of each individual contains a complete list of indices representing an output fuzzy set corresponding to each rule. Computing the fitness of each individual involves evaluating the FRBM on the training data. Fitness is defined as the inverse of the root mean square error between the defuzzified (numerical) output of the FRBM and the output of the training set.

5.10.2.   Problem description

The case study is performed on the same hydrodynamic model of the hypothetical estuary (Figure 5.6) used in §5.9. The model is again based on the finite difference solution of the 1-D full dynamic de Saint Venant equations (equations (5.7) and (5.8)).

As in the previous case study, the model uses the upstream flow and the downstream depth as boundary conditions. The upstream boundary data is different from the previous case study. It is taken from hourly flow data of a similar size river (Figure 5.15). The downstream boundary data is generated assuming a sinusoidal tidal wave with a period of 12 h and amplitude of 2.5 m (Figure 5.16). The downstream end is assumed to have a depth of 15 m at mean sea level.

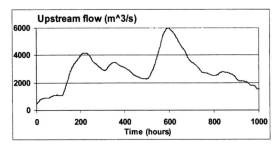

Figure 5.15. Upstream boundary condition

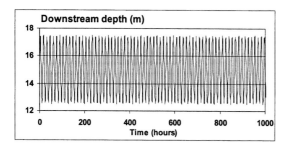

Figure 5.16. Downstream boundary condition

Similar to the previous case study, the discrepancy in water level prediction is introduced artificially using the following two numerical solution procedures. The first solution procedure uses a constant time step (200 s) for the whole river reach throughout the simulation period, which has a possibility of amplitude and phase errors for Courant numbers different from unity. The second solution procedure uses varying time steps depending on the state variables in the river, trying to maintain the average Courant number close to unity, in which case the solution obtained is close to the analytical solution of the model equations.

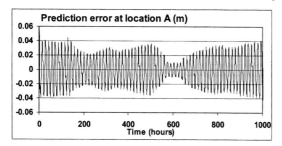

Figure 5.17. Errors in depth computation at location A

The model is run under both computational procedures for a simulation period of 1000 h. Point A, which lies at a cross-section halfway between the two boundaries of the reach, is selected as a point of analysis (Figure 5.6). The point corresponds to grid point 5 in Figure 5.7. The error in water level prediction at point A, which is taken here as the discrepancy between the predictions made using the two solution procedures, is shown in Figure 5.17.

5.10.3.   Application of the proposed method

From the above simulation, there are boundary data and prediction errors for 1000 h. The proposed geno-fuzzy method is applied step by step to this problem.

The first step is to select the best state variables for the IF part of the fuzzy rules. The upstream discharge and downstream water level seem appropriate here. One reason is that the error pattern appears to be related to these variables. Comparing Figure 5.15 and Figure 5.17, it can be noticed that error magnitudes are higher for lower discharges and vice versa. Similarly, comparing Figure 5.16 and Figure 5.17, it is apparent that the error time series shows a periodic behaviour obviously induced by the downstream tidal boundary condition. Another reason is that since both are the boundary conditions for the prediction model, they are readily available.

The second step is to select a proper lead-time between the selected variables and the model errors. Obviously there is some time involved for information to propagate from either boundary to the selected point of analysis, point A. For problems like this, the time of information propagation can be computed using the equations of wave speed or by using the method of characteristics; see for example Abbott (1966). However such techniques are not generic to every problem. Therefore AMI is used here. The AMI analysis revealed that the effect of the upstream discharge reaches point A in 1.5 h whereas that of the downstream tide reaches in 5 h.

The third step is to set the desired fuzzy sets and their corresponding membership functions on the range of the data representing the IF part of the rules (the upstream discharge before 1.5 h and the downstream tide before 5 h) and the THEN part of the rules (the prediction error at point A). Figure 5.18 shows five fuzzy sets ranging from VERY LOW to VERY HIGH constructed on the downstream water level data. In Figure 5.19, three fuzzy sets ranging from LOW to HIGH are constructed on the upstream discharge data. Figure 5.20 shows the membership functions of five fuzzy sets on the range of the error time series.

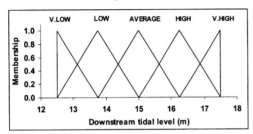

Figure 5.18. Fuzzy membership functions of the downstream tidal level

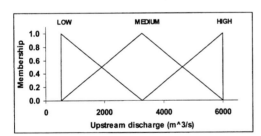

Figure 5.19. Fuzzy membership functions of the upstream discharge

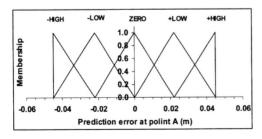

Figure 5.20. Desired membership functions on the expected prediction error

According to Figure 5.18 and Figure 5.19 there will be 5x3=15 rules since there are as many combinations of the premises such as:

"IF upstream discharge is LOW AND downstream level is VERY LOW, THEN prediction error is ... ".

Now that the premises (IF part) of the rules are set, the fourth step is to establish the consequence (THEN part) of each rule. The consequence of each rule will be one of the fuzzy sets on the prediction errors. The criteria is that all the rules combined must be able to represent the error landscape with reference to the two selected premises as well as possible. The genetic algorithm is applied with an initial population of 10 with the search tested up to 90 generations. In addition to the standard bit mutation that involves switching bits at random, creep-mutation that involves nudging random bit strings up or down.

Results

Figure 5.21 shows 15 rules generated from the training data using the proposed technique. The results present linguistic characteristics of the prediction uncertainty in the form of the relationship between boundary data and prediction errors. For instance, for high upstream discharges, the prediction errors are minimal. It also shows that model errors fall in the ZERO set when the downstream tide levels are at sea level (AVERAGE), whereas the model performs badly in case of extreme high and low tidal conditions. The results can be visually confirmed by observing Figure 5.15, Figure 5.16 and Figure 5.17.

| | | UPSTREAM DISCHARGE | | |
		LOW	MEDIUM	HIGH
DOWNSTREAM TIDE LEVEL	**V. LOW**	+HIGH	+LOW	+LOW
	LOW	+LOW	+LOW	ZERO
	AVERAGE	ZERO	ZERO	ZERO
	HIGH	-LOW	-LOW	ZERO
	V. HIGH	-HIGH	-LOW	-LOW

Figure 5.21. Model performance in terms of linguistic fuzzy descriptions

5.10.4.   Conclusions and discussion

The possibility of using an adaptive fuzzy rule-based structure to characterize the prediction uncertainty of a model is demonstrated. The prediction accuracy is mapped with respect to selected state variables, in this particular case, the boundary conditions.

One major advantage of this approach is that it deals with the overall prediction uncertainty since it works on the gap between the model prediction and corresponding observation. The other advantage is that the analysis results are presented in a linguistic form thus enabling the extraction of high-level knowledge from the historical performance of a prediction model. However, as might be expected, numerical precision will be lost in the transformation from thousands of numerical data to tens of fuzzy rules. A successful application of this approach depends on the proper utilization of physical insight, AMI analysis, the genetic algorithm and fuzzy set theory.

5.11. Case study: Application to flood routing on the Wye River

This case study demonstrates application of complementary modelling with the use of observed data in a river flood-forecasting problem. Also forecasts using a pure ANN model, a physically based model and a complementary modelling approach involving both models are compared. Information theory is used to detect the presence and time dynamics of residual information in errors, and to select input data to the ANN model. Unlike the case study on the hypothetical estuary, this case study is done using observed data and hence demonstrates the complementary modelling approach in a real-life problem. The material is adapted from Abebe & Price (2004).

5.11.1.   Introduction

The seriousness of river flooding in a number of countries highlights the need for better methods of predicting and forecasting high discharges. Most river floods are the result of heavy rainfall or snowmelt. When the generating rainfall or snowmelt is located in the headwaters of a river catchment then there are opportunities to base forecasts of flood discharges in downstream reaches on the time series of observed discharges at an upstream gauging station. This is particularly the case in long rivers such as the Mississippi and the Yangtse. The task of tracing the development of the flood from the upstream gauging site to points downstream is known as flood routing. Besides the use of sophisticated 1-D or 2-D computational hydraulic models based on solutions of the complete Saint Venant equations, say, there have been many simpler models proposed and applied to flood routing problems. Some of these methods are based on very simple physical concepts such as mass conservation and storage relationships, for example, the well-known Muskingum method; see McCarthy (1938). Other methods, although similar to the Muskingum method, are based on a better conceptualisation of the physical processes; see, for example, Cunge (1969) and a study by Price (1973) that led to the development of the Variable Parameter Muskingum-Cunge method; see also Price (1985).

The advantage of simpler flood routing methods is that they are often less demanding on data and are suited to real-time applications where faster simulations are needed. The danger that accompanies simplicity is the neglect of certain physical processes that may be critical in particular situations. This is certainly the case, for example, with flood routing methods that do not require a downstream boundary condition. Another difficulty with the simpler methods is that they trivialize the very complex processes that occur in long rivers whose lower reaches have large flood plains that can be extensively inundated during extreme events. Even if the additional flows feeding the river downstream of the gauging site are known accurately, the transformation of an upstream discharge time series to a corresponding time series downstream is subject in mathematical terms to sizeable non-linearity. Consequently, it is highly likely that a physically based routing model will have significant residual errors in its predictions of the downstream discharge in complex reaches.

5.11.2.   The routing model

The physically based model used in this case study is a particular form of the 1-D diffusive wave equation in which the second order derivative term is with respect to space and time rather than space alone as a strict diffusive term:

$$\frac{\partial Q}{\partial t} + c(Q)\frac{\partial Q}{\partial x} + c(Q)\frac{\partial}{\partial x}\left(\frac{a(Q)}{c^2(Q)}\frac{\partial Q}{\partial t}\right) = c(Q)q - c(Q)\frac{\partial}{\partial t}\left(\frac{Qq}{2gAs_0}\right) \qquad (5.9)$$

where Q is the discharge, $c(Q)$ is the convection speed, $a(Q)$ is an attenuation parameter, q is the lateral inflow, g is the acceleration due to gravity, A is the flow cross-sectional area and s_o is the longitudinal slope of the river. This equation was first derived by Price (1973) and is the basis of the Variable Parameter Muskingum-Cunge method.

Equation (5.9) is a consistent approximation of the full 1-D Saint Venant equations and has a form similar to a convection-diffusion equation with simplified wave speed and diffusion coefficients. Strictly, the non-linear nature of a river with extensive flood plains demands that c and a are both functions of discharge depending on the nature of the river between the upstream and downstream limits of the model. These functions can be deduced from data for the river. The numerical model based on equation (5.9) has the advantage that it can be solved using a straightforward explicit finite difference scheme that does not need a downstream boundary condition; see Price (1985). Although the non-linear equation above is justified in terms of its derivation, this case study focuses on the linearised form of the equation that assumes c and d are independent of discharge:

$$\frac{\partial Q}{\partial t} + c\frac{\partial Q}{\partial x} + d\frac{\partial^2 Q}{\partial x \partial t} = cq \tag{5.10}$$

where d is a diffusion coefficient, which is equivalent to a/c^2 in equation (5.9).

These flow equations can be schematised into a finite difference scheme that has the form:

$$C_0 Q_{j+1}^{n+1} + C_1 Q_{j+1}^n + C_2 Q_j^{n+1} + C_3 Q_j^n + C_4 = 0 \tag{5.11}$$

where C_0 to C_4 are coefficients, and j and n are the space and time grid levels respectively.

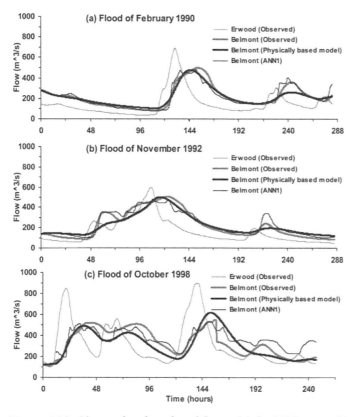

Figure 5.22. Observed and predicted flows: (a) the 1990 event, (b) the 1992 event, and (c) the 1998 event

5.11.3. Study area and data

The case study was focused on a 70-km reach of the River Wye, UK, between gauging stations at Erwood and Belmont. The interest in this reach was that it has extensive unembanked flood plains that are inundated for the more extreme events and which can lead to significant attenuation of flood peaks recorded at Erwood. Data from the floods observed in February 1990, November 1992 and October 1998, hereafter referred to as the 1990, 1992 and 1998 events, respectively, were used. Looking at the base flow parts of the hydrographs

(Figure 5.22(a) and (b)), it is evident that there was some lateral inflow along the river reach. The amount is generally small compared with the significant flood discharges. The 1998 event (Figure 5.22(c)) should be treated with caution in that there appears to have been a monitor failure during the last part of the event.

Preliminary information content analysis

At first, the AMI measures of the flow at Belmont with itself and with flow at Erwood were computed at varying lag times (Figure 5.23). The former shows the information within flow data at Belmont separated in time. The latter is a measure of the predictability of the flow at Belmont using flows observed at Erwood. The lag time corresponding to the maximum AMI between Erwood and Belmont is between 11 and 12 h, which can be interpreted as the average time of flood propagation along the reach. It also indicates that forecasts at Belmont can be made using data at Erwood up to this horizon with a consistent accuracy. The deduced information was used to estimate the range of possible values for the convection speed for the physically based model. The model uses a time step of 15 minutes and a spatial resolution of 2 km. The convection speed and attenuation parameters were fixed by automatic calibration. The bounds of the convection speed were set on the basis of the information travel speed along the reach. A small, constant lateral inflow was assumed.

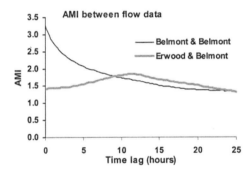

Figure 5.23. AMI of flow at Belmont with itself and with flow at Erwood

5.11.4.   Routing with ANN model

An RBF type of ANN routing model was also developed. This model, labelled here as ANN1, was developed to evaluate the relative accuracy of the physically based model. It uses as input only the discharges at Erwood. Three input nodes with a lag time of 11-12 h were used. These input nodes were selected on the basis of the lag time corresponding to the maximum AMI between the flows at Erwood and Belmont (see Figure 5.23). The network was trained with the 1990 event and applied to the other two events.

Table 5.3. RMSE of flow forecasts at Belmont in m^3/s (only upstream data is used)

Model type (1)	1990 (2)	1992 (3)	1998 (4)
Physically based model	31.16	42.40	78.61
ANN1	32.56	37.12	78.31

The resulting hydrographs at Belmont computed by the physically based model and ANN1 are shown in Figure 5.22. The RMSE between the observed and computed discharges at Belmont are shown in Table 5.3. Despite the difference in the shape of the hydrographs computed by the two models, the corresponding RMSE measures showed comparative values for the three events. The accuracy of both models for the 1998 event in the last part of the hydrograph was particularly poor, perhaps because of the data.

Residual information content analysis

A second AMI analysis was carried out to detect the predictability of the residuals of the physically based model. A number of time series were considered for the analysis: the discharge at Erwood, the first, second, and third derivatives of discharge with respect to time and space, the calculated discharge at Belmont, the observed discharge at Belmont, and the error of the physically based model itself.

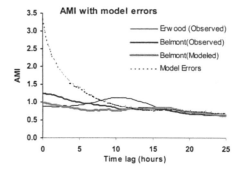

Figure 5.24. AMI between errors and principal state variables

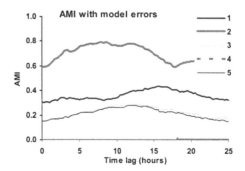

Figure 5.25. AMI between errors and derivatives of discharge at Erwood, where: 1=dQ/dt, 2=dQ/dx, 3=d²Q/dt², 4=d²Q/dx², 5=d³Q/dt³

Figure 5.24 shows the AMI that the upstream discharge and the measured and predicted downstream discharges share with the errors of the physically based model. The figure shows that the most dominant level of information derives from the model errors themselves at the previous time steps. Its magnitude decreases from about 2.5 at a lag time of 15 minutes backwards in time. The figure also shows that fairly high levels of information are available from the upstream discharge observed between 10-12 h before the event and the downstream discharge observed just before the event. Apparently, the fact that the discharge at Erwood and the forecast errors at Belmont share a maximum level of information at a lag time of 10-12 h shows that the physically based model in its structure or parameters has not used all the information available in the upstream discharge in predicting the downstream discharge. The high AMI values in the immediate past errors and the consistent decrease with lag time indicate that if previous errors are part of the input data to forecast future errors, the quality of the forecast will be high for shorter horizons and decreases with increasing horizon.

Figure 5.25 shows the AMI between the errors and first and higher order derivatives of the upstream discharge with respect to space and time. These are plotted at various lag times. Note that two graphs (Figure 5.24 and Figure 5.25) are used to display the AMI for different variables partly for clarity but also because a different vertical scale is needed to magnify the relatively faint levels of AMI for particular variables. Comparing the two figures, it can be seen that derivatives are less important than other variables for the error prediction model. In particular, Figure 5.25 shows that the higher order derivative terms are not related to the residuals.

5.11.5.   Error forecasting with ANN model

The results of the residual AMI analysis can now be used to select the best-related input data for the error forecast model. An ANN model of the RBF type was used to forecast the errors of the physically based model. This network, labelled here as ANN2, was constructed using three input nodes: two nodes for previous errors and one node for the upstream discharge. The selection of input nodes was again based on the AMI values shown in Figure 5.24.

The 1990 event was used to train ANN2. The resulting ANN model was applied to forecast the errors of the 1992 and 1998 events. The prediction of the errors was carried out at horizons varying from 15 min to 4 h. The input data consisted of the errors taken immediately before the event, and the upstream discharge before 10 h (40 time steps). Therefore the errors are the constraints that limit an increase in the forecast horizon since the information content fades with the forecast horizon. If the lead-time extends beyond 10 h, the information from the discharge at Erwood also decreases. In that case, the physically based model itself will not have sufficient information to make the flow forecast. One remark for this case problem is that even better results would have been expected if other catchment data such as precipitation and groundwater level were used in the error forecast. It is possible that these data are related to the lateral inflow and the baseflow of the river.

5.11.6.   ANN routing with downstream information

A third ANN model, ANN3, was constructed to forecast the flow at Belmont. This ANN uses information both at Erwood and Belmont. ANN3 was used to make a comparison with the combined performance of the physically based model and the error forecasting ANN model. ANN3 is composed of three input nodes: two nodes from flow at Belmont and node from flow at Erwood. The input data and corresponding lag time were selected according to the preliminary AMI analysis shown in Figure 5.23. Since ANN3 uses downstream data, its accuracy depends on the forecast horizon. The 1990 event was used to train ANN3.

Results

Table 5.4. RMSE of forecasts at Belmont in m^3/s (using both upstream and downstream data)

Horizon (1)	Physically based model and ANN2			ANN3		
	1990 (2)	1992 (3)	1998 (4)	1990 (5)	1992 (6)	1998 (7)
15 min	2.18	6.10	42.78	5.22	8.81	22.80
30 min	2.40	7.92	44.28	5.65	9.02	24.26
1 h	3.12	10.20	45.76	5.92	10.83	28.17
2 h	3.72	14.70	49.00	8.10	16.30	36.58
3 h	5.83	19.64	53.63	9.60	20.41	41.87
4 h	8.05	25.55	59.43	12.83	25.61	50.62

Table 5.4 indicates the RMSE between observed and computed discharges at Belmont for the physically based model along with ANN2 at different horizons. The table also shows the RMSE for ANN3. Comparing the corresponding values in Table 5.3 and Table 5.4, it is evident that including additional downstream information by the complementary ANN model (ANN2) reduces the RMSE thus improving the forecast accuracy. ANN3 also forecasts discharges more accurately than ANN1 showing that, in all the cases, additional information significantly improves the forecast accuracy.

In Table 5.4, a comparison of columns 2, 3 and 4 with columns 5, 6 and 7 shows that for the training event (1990) and the first verification event (1992), using ANN2 along with the physically based model gives better forecasts than ANN3. For the 1998 event, ANN3 performed better. However, the suspicious nature of the data for the 1998 event prevents any definite conclusion.

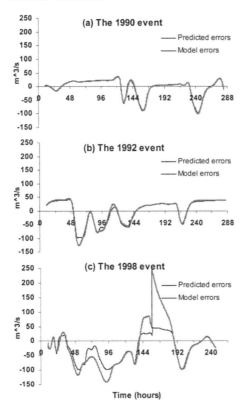

Figure 5.26. Model errors and predicted errors in flow forecast at Belmont: (a) the 1990 event, (b) the 1992 event, and (c) the 1998 event

Figure 5.26 shows the residuals of the physically based model and the corresponding errors forecasted by ANN2. In particular, Figure 5.26(a) and (b) show a close reproduction of the errors for the 1990 and 1992 events. For the 1998 event, Figure 5.26(c) shows that there are parts of the routing model error not adequately captured by ANN2. The error prediction model is trained on the errors from the simulation of the 1990 event, which range from -100 m^3/s to $+35$ m^3/s. The magnitudes of the positive errors of the 1998 event are simply outside the range of the training data, which makes it an extrapolation problem. The solution to this problem is to use more diverse training data including data with large magnitude. Such an approach will help to train the ANN to handle errors with large magnitudes. However, the particular peak level simulation errors on the seventh day (after 144 h) of this event are caused by a monitor defect. This is evident by looking at the unusually steep drop in the flow observed at Belmont (Figure 5.22c) at about 155 h. Even in this situation, the error model has predicted part of the residual errors.

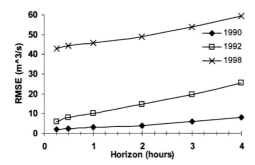

Figure 5.27. RMSE of forecast at Belmont after applying the error forecast model

In Figure 5.27, the RMSE of the forecasts at Belmont at different horizons is shown. The forecasts improve for shorter horizons. The forecast accuracy of the 1998 event worsens faster than the other two events. This is because RMSE grows faster when large values are involved. Figure 5.28 shows the gain of accuracy versus forecast horizon. The gain of accuracy was calculated as the difference between the RMSE of the forecasts at Belmont by the physically based model with and without error forecasting. The gain of accuracy is very significant for short horizons and decreases, as one would expect, for larger horizons. There is a marked reduction in the errors even for the 1998 event for which the observed data is known to be suspect.

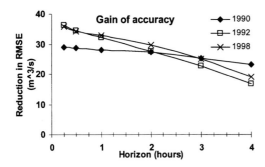

Figure 5.28. Reduction in the RMSE of forecasts at Belmont as a result of error forecast

The gain of accuracy in discharge forecasts is even more important when translated into water level. For instance, for the 1992 event corresponding to time 100 h, there is a gain of accuracy in the flow forecast amounting to 50 m^3/s (Figure 5.26(b)) corresponding to an observed flow of 450 m^3/s (Figure 5.22(b)). Using the stage-discharge relation of the Wye River at Belmont, this is equivalent to 40.75 cm of improvement corresponding to a water level of 4.9 m, which is a significant improvement in terms of flood inundation. The water level on the Wye is in the range of 2 to 4 m at Belmont.

5.11.7.   Conclusions and discussion

The study indicated that AMI analysis helps to study how data are related to the errors, to obtain information about which data can be used to recover errors, and to obtain an insight into the time dynamics between the data and model errors. It has been shown that errors of the flood routing model share varying degrees of information with the input data, the output data and the state variables of the model at previous time steps. The particular time series that share maximum amount of information with the residual errors have been identified and used to construct a complementary ANN model that forecasts the expected errors of the routing model. The AMI analysis also helped in the estimation of the bounds of the convective speed for the physically based model and the optimal lead times for the ANN routing models.

The results indicated that forecasting errors with a data-driven model considerably improves the forecast accuracy. The possibility of forecasting errors with comparatively simple empirical model eases the need for developing a complicated physically based routing model. Complex, time and data demanding, but relatively less important processes, can be ignored when setting up the routing model. After a preliminary forecast is made, results can be refined using models that can learn from data.

Also the analysis of model errors indicates possible ways to make further improvements to the routing model. For instance, the routing model considers the lateral inflow as a constant in space and time. However, it is known that it depends on factors such as the precipitation in the catchment area (619 km^2) contributing exclusively to the reach and any interaction of the river with ground water. Generally, lateral inflow does not vary significantly over a period of several hours due to the retention effect of the catchment and the relatively slow dynamics of the interaction with ground water. Figure 5.24 shows that residual errors have high AMI values within 5 h. It can therefore be argued that the way in which the lateral inflow is handled by the routing model makes a significant contribution to the prediction errors.

One possibility to further improve the complementary model is to investigate the importance of other data to develop a better complementary model. For instance, even better results may be obtained if other catchment data such as precipitation and groundwater level were available and used in the complementary model. Such data are related to the lateral inflow and the baseflow of the river.

5.12. Case study: Application to a conceptual rainfall-runoff model

This case study presents a demonstration of the complementary modelling approach to runoff prediction using a conceptual rainfall-runoff model of the Sieve basin in Tuscany, Italy. The data of the Sieve basin is one of the most highly used data in the literature on rainfall-runoff modelling. The accuracy of the forecasts is also compared with those reported in previous research. Unlike the case studies on the hypothetical estuary and River Wye, a neural network model of the multi-layer perceptron type is used as a complementary model. Multiple ANN models are trained to forecast the residuals of the conceptual model for a horizon of 1–6 h. This material is adapted from Abebe & Price (2003b).

5.12.1.   Conceptual rainfall-runoff models

Compared to detailed physically based models, conceptual rainfall-runoff models are very convenient in that they are less demanding on structural data from the physical domain, yet maintain an acceptable forecast accuracy and reflect the essential aspects of the physics. These models generally have two major components: one representing the soil water balance and the other the evolution of the flow to the outlet of the basin. The concepts used to represent these two components are the basis for the differences between the conceptual models developed over the years. Franchini & Pacciani (1991) compared various conceptual rainfall–runoff models that use different structures to represent the involved processes. Their results showed the variation in the performance when applied to the same data. The structure that conceptual models adopt to handle different aspects of the process varies and, consequently, is a factor in the success of the modelling.

Typical characteristics of conceptual models include variables representing average values over the entire basin, or parts of the basin, and model parameters that cannot be assessed from field data but have to be obtained through calibration. However, interdependence between the parameters, the danger of stretching them beyond their theoretically known bounds when minimizing residuals, and the absence of a unique best set of parameters impose parameter uncertainty. Uhlenbrook *et al.* (1999) emphasised the importance of parameter uncertainty compared to different model variants. They argued that predicted discharges should be presented as ranges instead of single values, since different parameter sets yield widely varying responses during application. Approaches by other authors suggest the development of different model variants for low- and high-flow predictions and then combining them (see for example Shamseldin *et al.*, 1999). The use of poor quality data for calibration also affects the value of parameters. Mroczkowski *et al.* (1997) stressed the importance of incorporating data with a changed response and using data other than stream flow in the validation of conceptual models to obtain more representative parameters and model structure.

In recent years, there has been the growing application of data-driven models in rainfall–runoff modelling: ANN models are examples of widely used data-driven models. Data-driven models are particularly suitable in rainfall-runoff modelling mainly because they use time series data and not structural data from the physical domain. Recent examples of the application of ANN models to rainfall–runoff modelling can be referred to in Minns & Hall (1996), Dawson & Wilby (1998), Rajurkar *et al.* (2002), Campolo *et al.* (2003) and Solomatine & Dulal (2003).

5.12.2.   The ADM model

The conceptual rainfall–runoff model used here is ADM ("a distributed model" - see Franchini, 1996 for a detailed description). The model consists of components for soil water balance and transfer of the flow to the basin outlet. The soil water balance component expresses the balance between the moisture content of the soil and the incoming (precipitation) and outgoing (evaporation, surface runoff, interflow and baseflow) quantities. Surface runoff is calculated using the soil storage capacity curve method applied in the XINANJIANG model (Zhao *et al.*, 1980). The transfer component computes the flow along the hill slopes and the flow in the streams using a convection–diffusion equation, with a constant lateral inflow term in the former case, with the same approach used in the ARNO model (Todini, 1996). The ADM has 11 parameters that are usually calibrated.

Study area and data

The study area is the Sieve basin, which is a sub-basin of the Arno basin in Tuscany, Italy. The basin has four sub-basins and a total area of 822 km^2, a considerable part of which is hilly and mountainous. A three-month, hourly data set (December 1959–February 1960) was used. The data comprise hourly water levels at Forcancina, hourly mean precipitation and hourly evaporation estimates. The stream flow was calculated from the hourly water levels using a rating curve. The mean areal precipitation was calculated by the Theissen polygon method using data from 11 rainfall stations. The choice of the study area and the data is motivated by the fact it has been used as reference case to test and compare several hydrological models and model calibration algorithms in Franchini & Pacciani (1991); Franchini *et al.* (1996) and Franchini *et al.* (1998).

To study how the rainfall and runoff are related, the coefficient of linear correlation and the AMI between rainfall and runoff data were computed up to a lag time of 10 days, as shown in Figure 5.29 and Figure 5.30, respectively. Both graphs, the former from a strictly linear measure and the latter independent of any function, show similar trends in the relationship, indicating that the response of the basin is largely linear. However, the presence of sizeable non-linearity in particular ranges of the data cannot be ruled out, since both techniques were applied on the whole data record.

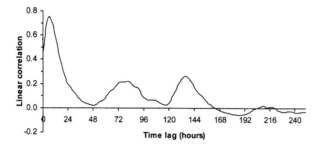

Figure 5.29. Linear correlation between rainfall and runoff

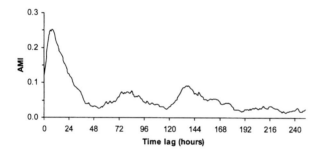

Figure 5.30. AMI between rainfall and runoff

Both measures (Figure 5.29 and Figure 5.30) also indicate that there are three distinct peaks in the relationship between rainfall and runoff. There is one immediate response between 6 and 8 h and two delayed responses: 3–3.5 days and 5.5–6 days. This does not come as a surprise, since the basin can respond to precipitation event in the form of direct runoff, interflow and baseflow. However, interpretation of the basin response into these components needs more information than indicated by the above measures. Nevertheless, the rainfall has a considerable contribution to the runoff for at least 6 days. Similar interpretation was made in Young (2001) regarding the presence of responses with distinct delays.

ADM model performance

The ADM model was applied to the data from the Sieve basin as a lumped model over the entire basin. A time step of 1 h was used. The 11 parameters of the model were determined by calibration using the data of December 1959 (the first one-third of the data). Calibration was done using four optimisation algorithms, namely, genetic algorithm, adaptive cluster descent, adaptive cluster covering with local search and controlled random search using the optimisation package GLOBE described in Solomatine (1999). The best parameter set was selected based on the performance of the model on the remaining two months of the data. The *RMSE* between model and observation for the calibration and validation sets are 19.15 and 25.3 m³/s, respectively. The corresponding values of the coefficient of model efficiency, *CE* (after Nash & Sutcliffe, 1970) are 0.889 and 0.887, respectively. For the whole time series, the error in runoff prediction varies between –200 and +125 m³/s, the *RMSE* being 25.2 m³/s.

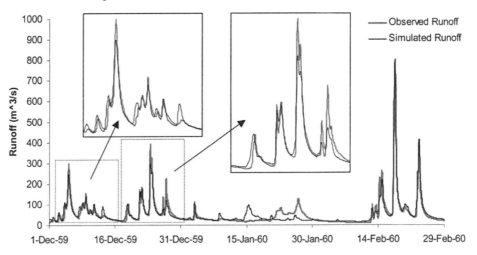

Figure 5.31. Observed and simulated flows at Forcancina (Dec 1959 – Feb 1960)

As can be seen in Figure 5.31, the model shows a reasonable reproduction of the observed runoff for most of the data. However, there are observable discrepancies in some portions of the hydrograph. For instance, in the period between 12 and 29 January 1960 (Figure 5.31), there is an evident difference between the observed and modelled runoff. Similar behaviour over the same segment of the data is seen in a comparison of seven different conceptual rainfall–runoff models (namely, STANFORD IV, SACRAMENTO, TANK, APIC, SSARR, XINANJIANG and ARNO models) conducted by Franchini & Pacciani (1991); also the TOPMODEL (Franchini *et al.*, 1996). The climate of the study area and the fact that this portion of the data set lies in the month of January suggest that such a response is possibly the effect of snow, which is not considered in the conceptual model. The overall performance of the ADM model shows the same order of magnitude of the coefficient of efficiency, *CE*, as that of the other conceptual models applied to this basin. (See Table 1 and Figure 4 in Franchini *et al.*, 1996. A direct comparison of *CE* on the validation set is not possible since one extra month of data, March 1960, which was not available for this study, was used in that article). Also care should be taken in the interpretation of the coefficient of efficiency. A study by Hall (2001) has shown that the coefficient of efficiency is less sensitive to the underestimation of runoff volumes and it tends to fall below zero if there is a strong bias

5.12.3.   Complementary model development

AMI Analysis

The AMI between the residual error of the conceptual model and the observed rainfall and runoff data, as well as the error itself up to a lag time of 25 h are plotted in Figure 5.32. It shows that the model errors share a maximum level of information with the observed rainfall at a lag time of between 4 and 5 h and with the observed runoff just before the event. The highest level of information about errors is available in the previous errors. This level of information decreases with increasing lag time.

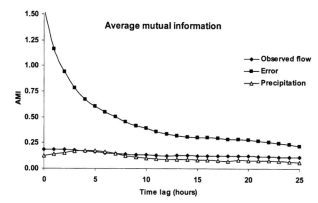

Figure 5.32. AMI between error and other time series

The presence of a relationship between errors and other time series suggests the presence of a systematic process causing errors and therefore the possibility of capturing them with a complementary model. At this point there is not enough evidence to speculate on what the actual cause of the errors could be, but it is probably a combination of misrepresented processes, incorrectly defined model parameters, structure and equations, and the quality of data.

The complementary model

An ANN model of the MLP type is used as a complementary model in this case study. A three-layered network with five nodes in the input layer and one node in the output layer (see Figure 5.33) is used. Two nodes from the past rainfall, one node from the antecedent flow and two nodes from the model errors at previous time steps are used to predict the errors. The sigmoid function is used in the hidden nodes, whereas the linear function is used in the output nodes.

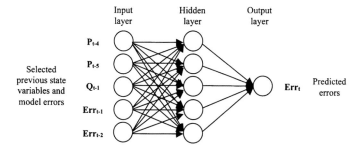

Figure 5.33. Neural network structure

A problem associated with using ANN models to complement conceptual models is over-parameterisation that can result from using two over-parameterised models together. However, this is not as bad as it seems, since the two models are developed one after the other and not simultaneously. Also, care is taken to avoid an oversized network by limiting the number of nodes in the hidden layer to five. Moreover, a separate ANN model is developed for each horizon. This means that the number of nodes in the output layer is only one in all cases, despite the possibility of using one ANN with multiple outputs corresponding to each horizon.

The tests are conducted in three stages:

At first, the relative performance of the ADM model, a naïve model and ANN error prediction models (set up on a 1-h lead-time basis) are tested. Here, the naïve model is the one that assumes the forecast runoff to be equal to the latest available runoff record. This model can sometimes, particularly for shorter lead-times, reproduce the runoff closely, though it has an obvious problem of phase errors. For the ANN model, the available data record is used in three variants. In the first variant (ANN I) the whole data set (2160 examples) is used for training. In the second variant (ANN II), the first part of the data set (1160 examples) is used to train the network, while the second part (the remaining 1000 examples) is used for verification. For the third variant (ANN III), the first part of the data set is used to verify the network trained using the second part. The accuracy of these tests is shown in Table 5.5.

Table 5.5. Comparison of the performance of different models on different data segments

Model variant	Data set	RMSE (m^3/s)	CE
ADM model	First part	20.0	0.841
	Second part	26.7	0.907
Naïve model	First part	8.0	0.975
	Second part	12.1	0.981
ANN I + ADM	All data (training)	4.8	0.995
ANN II + ADM	First part (training)	4.3	0.993
	Second part (verification)	6.2	0.995
ANN III + ADM	Second part (training)	4.9	0.997
	First part (verification)	4.9	0.990

The second stage is mainly intended to evaluate the efficiency of the ANN error forecast model for horizons of 1–6 h. For this purpose, the first part of the data set is used for training and the second part is used for verification in all the tests of the second stage; in other words the second variant (ANN II) is used. This essentially means that six neural networks (one for every horizon) have to be trained with different input file configurations. This is needed to ensure the availability of the data in operational forecast mode. The network structure shown in Figure 5.33 is maintained for all horizons. Table 5.6 shows the data composing the input nodes corresponding to various horizons.

Table 5.6. Data organization for the complementary ANN model

Horizon (h)	Input nodes	Output node
1	P_{t-4}, P_{t-5}, Q_{t-1}, Err_{t-1}, Err_{t-2}	Err_t
2	P_{t-4}, P_{t-5}, Q_{t-2}, Err_{t-2}, Err_{t-3}	Err_t
3	P_{t-4}, P_{t-5}, Q_{t-3}, Err_{t-3}, Err_{t-4}	Err_t
4	P_{t-4}, P_{t-5}, Q_{t-4}, Err_{t-4}, Err_{t-5}	Err_t
5	P_{t-5}, P_{t-6}, Q_{t-5}, Err_{t-5}, Err_{t-6}	Err_t
6	P_{t-6}, P_{t-7}, Q_{t-6}, Err_{t-6}, Err_{t-7}	Err_t

In the third stage, the ADM model is run in forecast mode to make runoff forecasts up to 6 h in advance. This is done assuming the future precipitation to be equal to the last observed value. The experiment in the second stage is repeated on the forecasts made by the ADM model. The *RMSE* and *CE* of the ADM forecasts are shown in Table 5.7 without and with updating by ANN model.

Table 5.7. Performance of the complementary model (applied on top of ADM) on the verification set (second part of the data). Training is done on the first part in all cases.

Horizon (h)	Naïve model		ANN+ADM (prediction mode)		ADM model (forecast mode)		ANN+ADM (forecast mode)	
	RMSE (m^3/s)	CE	RMSE (m^3/s)	CE	RMSE (m^3/s)	CE	RMSE (m^3/s)	CE
1	12.1	0.981	6.2	0.995	27.04	0.905	7.54	0.993
2	23.3	0.929	12.3	0.980	27.74	0.900	15.22	0.970
3	33.5	0.854	18.0	0.958	30.19	0.881	25.34	0.916
4	42.7	0.763	21.2	0.942	35.12	0.840	32.05	0.866
5	50.8	0.664	22.7	0.933	42.37	0.766	40.42	0.787
6	58.1	0.561	23.9	0.926	51.36	0.657	46.61	0.717

Results

Figure 5.34 shows the verification performance on the second part of the data set using the neural network trained on the first part (ANN II). Similarly Figure 5.35 shows the verification results on the first part of the data set using the network trained on the first part (ANN III). In both cases, a horizon of 1 h is used. Both cases indicate that the errors could be considerably minimised by the complementary model. Even the simulation errors corresponding to the segment of the hydrograph between 12 and 29 January 1960 in Figure 5.31, which is presumably caused by snow, is well reproduced by the complementary model.

The results of the first stage in Table 5.5 show the *RMSE* and *CE* between observed and predicted runoff using the ADM model, the naïve model, and error prediction model applied on top of the ADM model. All the three variants of ANN applied with the ADM model show that there is a major improvement in the prediction when error prediction models are applied compared to the conceptual model alone and the naïve model.

The results corresponding to the second and third stages are shown in Table 5.7. ANN II, even when applied 6 h in advance, was able to improve the performance of the ADM model operated in prediction mode, shown in Table 5.5. In forecast mode, the naïve model performed better for the first 3 h compared to the ADM. For longer than 3 h, ADM forecasts are better. In all the cases, when ANN II was applied on top of ADM forecasts, there was a consistent improvement in the forecast accuracy according to the *RMSE* and *CE* measures.

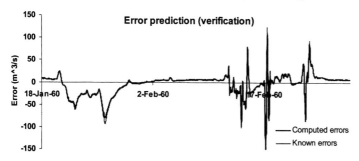

Figure 5.34. Error prediction on the second part of the data with 1 h horizon (network trained on first part)

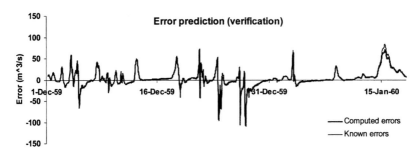

Figure 5.35. Error prediction on the first part of the data with 1 h horizon (network trained on second part)

5.12.4.   Conclusions and discussion

The AMI and correlation analyses showed that there are at least three distinct responses by the catchment. AMI analysis also helped to detect residual information in the errors of the conceptual model. The highest level of information about errors in runoff prediction is contained in the errors at previous time steps. The content of this information decreases with increasing lag time. This actually limits the feasible lead-time for the application of the complementary model. The data and the corresponding lag times that correspond with maximum levels of information have been identified and used to develop the complementary ANN models. Subsequently, the complementary models are used to forecast the errors up to 6 h in advance. As a result, the accuracy of the runoff forecasts has improved significantly.

The significant improvement achieved in runoff forecast using complementary modelling can be attributed to two causes. First, the data-driven model uses data that are not used by the conceptual model. For example, the conceptual model does not use the antecedent runoff whereas the complementary ANN model uses it as a direct input. The second factor is that the complementary model uses the forecast errors at previous time steps. It is a crucial piece of information, since it reflects the gap between the actual response of the basin and the response of the conceptual model. This is probably the reason why the ANN predicts the errors resulting from the suspected snowmelt. It uses antecedent model discrepancy and the antecedent flow. The use of antecedent model discrepancy is particularly important to manage errors resulting from slowly varying processes.

The results indicate that it would be of great benefit to apply error prediction models along with conceptual models to obtain better forecasts. It is a potentially good approach to incorporate the effect of processes that are not included in the conceptual model. Such a forecasting approach not only reduces forecast errors but also reduces the uncertainty by helping to bridge part of the gap between the conceptual model and the actual response of the basin.

The advantages of complementary modelling lie in the fact that, once a complementary model is established for a basin and its conceptual model, the complementary model is practically independent of the main model, which means there is no need to re-run the conceptual model. Also, the complementary model can incorporate input data with which the conceptual model is not structured to work. However, if the structure or parameters of the conceptual model are altered in some way, then its error pattern might change and the complementary model has to be re-trained to fit the error patterns of the altered conceptual model.

PART III. APPLICATION

CHAPTER 6. FLOW FORECASTING ON THE RHINE AND MEUSE RIVERS

This chapter illustrates the importance of complementing traditional forecast models with data-driven models in a practical problem of river flow forecasting. The methodologies discussed in the previous chapters of the thesis are applied to models that are developed to make 24 h in advance flow forecasts at three locations on the Rhine and Meuse, two major rivers in The Netherlands. The purpose of the forecasted flows is to serve as the riverside boundary conditions of the *Zeedelta* model, which is a hydrodynamic model that covers the delta area of the two rivers and part of the North Sea near the Dutch coast. The Zeedelta model is intended to forecast water currents as they affect navigation of shipping entering and leaving the Port of Rotterdam and points upstream. There is also a need to forecast the intrusion of salt water as it may affect the intakes for drinking water. Flow forecasts are to be made at three locations where the two rivers enter the delta. These locations are Hagestein and Tiel on the Rhine branches and Lith on the Meuse. The study area is shown in Figure 6.1.

Figure 6.1. The Rhine and Meuse rivers in The Netherlands

For both rivers, independent application of physically based and neural network flow routing approaches, both using the data at upstream measuring locations to forecast downstream flows, are compared with the complementary application of the two modelling techniques. The practical challenges involved in developing the forecast models for the two rivers are different and therefore the complementary modelling procedure is posed in different ways. For the Rhine River, updating is focused on the final output of the model. It is illustrated by incorporating an external process, namely, the tidal influence, indirectly using a complementary data-driven model. The term 'external' is used since the original model does not have provisions to model tidal influence. For the Meuse, the intermediate output of the primary model is updated. The possibility of incorporating information obtained from data for times when it is not available is illustrated. The study also compares the operational aspects and data needs of the two modelling techniques. Part of the material in this chapter is presented in Abebe *et al.* (2003).

6.1. The Rhine and Meuse Rivers in The Netherlands

The Dutch part of the Rhine River begins from the location Lobith near the border with Germany. This reach of the Rhine is called Upper-Rhine (Bovenrijn). The Upper-Rhine branches into the Waal and the Pannerden Canal (Pannerdens kanaal). The Pannerden Canal in turn branches at Westervoort near Arnhem into the Lower-Rhine (Nederrijn) and the IJssel. The Waal and the Lower-Rhine flow west into the North Sea whereas the IJssel flows north into Lake IJssel (see Figure 6.2). The river system has undergone continual changes both naturally and as a result of human intervention. During high flow rates at Lobith, the distribution of flow is approximately $2/3^{rd}$ for the Waal, $2/9^{th}$ for the Lower-Rhine and $1/9^{th}$ for the IJssel (see RIZA, 1993). During low flows, the distribution of flow into the branches is regulated by structures across the Lower-Rhine branch in addition to the natural river geometry. The highest flow recorded at Lobith is 12600 m^3/s in 1926. The lowest on record is 620 m^3/s in 1947, which is 20 times lower than the highest on record (Silva *et al.*, 2001).

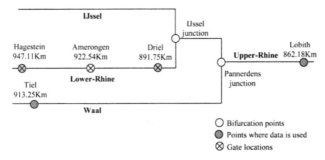

Figure 6.2. Schematic diagram of the Rhine branches

The Dutch part of the Meuse River unlike the Rhine is a continuous river reach that goes all the way from its entry to The Netherlands directly to the North Sea. Reports indicate that the historical maximum and minimum discharge records on the Meuse vary by a factor of 150. This variation in discharges is high compared to that of the Rhine since the reach of the Meuse has a number of tributaries with significant contributions to the flow (Silva *et al.*, 2001). The Meuse is also regulated by several gates in order to fulfil navigation requirements during low flows. As it is shown schematically in Figure 6.3, there are seven gates between Borgharen and Lith.

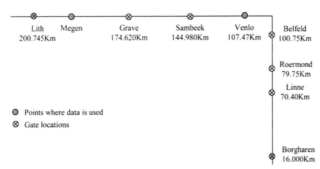

Figure 6.3. Schematic diagram of the Dutch part of the Meuse

Parts of the reach are also undergoing changes. An investigation by WL | Delft Hydraulics (2001) indicated that in the framework of the "Deltaplan Grote Rivieren (DGR)" project, dikes and levees have been constructed and/or raised at various locations along the reaches of the Grensmaas (Borgharen to Roermond) and Zandmaas (Roermond to Lith) since 1995. Within the framework of the project "Zandmass/Maasroute", three pilot dredging projects are or will be executed in the Zandmass. The report emphasizes the potential effect of the projects on the attenuation and travel-times of flood waves between Borgharen and Lith.

6.2. Preliminary data analysis

The main source of discharge and water level data used here is the DONAR database, which consists of 10-min and 24 h discharge and 1 h water level records at various locations on the Rhine and Meuse rivers. For the Rhine branches, data from January 1996 to September 2001 is used. For the Meuse, data recorded from January 1997 to September 2001 is used.

6.2.1. Data relationship

The relationship between data is analysed using AMI measures. The lag time corresponding to the maximum AMI between data at two locations is considered as the average information propagation time. The information propagation times are later used to estimate the wave celerity for physically based modelling and to establish proper input-output connection for data-driven modelling. AMI analysis requires the use of continuous records. Since the available data consist of gaps, continuous parts of the time series are used to carry out AMI computations. The AMI analysis is done before filling the gaps intentionally since the results of the AMI analysis are used in the process of filling the gaps in the data. The AMI measures are computed between different time series up to a lag time of 35 h as shown from Figure 6.4 to Figure 6.7.

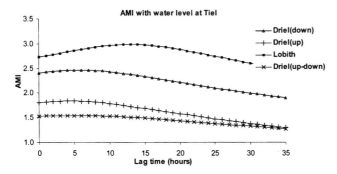

Figure 6.4. AMI with water level at Tiel

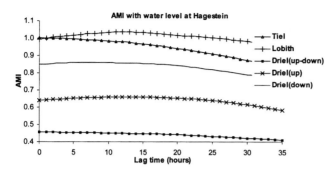

Figure 6.5. AMI with water level at Hagestein

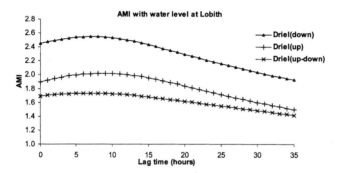

Figure 6.6. AMI with water level at Lobith

Figure 6.4 shows the AMI between the water level at different locations on the Rhine branches and the water level at Tiel at varying lag times. The figure shows that the water level at Lobith is particularly important since it is strongly related to the water level at Tiel. The maximum AMI corresponds to a lag time of 13 h, which indicates the average time for information to propagate from Lobith to Tiel. Figure 6.5 shows similar information travel time from Lobith to Hagestein. Figure 6.4, Figure 6.5 and Figure 6.6 show that at Driel the water level downstream is more relevant than upstream. This is a result of the influence of the gate at Driel. From Driel to Tiel the average travel time is 6 h. Since Driel and Tiel are not on the same branch, it has to be noted that there is no physical movement of water involved between the two locations. However, both branches receive information from Lobith.

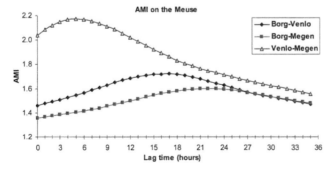

Figure 6.7. AMI analysis on flow data of the Meuse

The results of the AMI analysis on the flow data of the Meuse (Figure 6.7) show that the average travel time from Borgharen to Venlo is 16 h whereas from Borgharen to Megen it is 22 h. The average travel time for the reach from Venlo to Megen is 6 h. Since there is no adequate data at Lith, the travel time in the relatively shorter reach from Megen to Lith could not be computed. However, assuming the same information propagation speed as in the preceding reach, the travel time from Megen to Lith can be estimated as 2 h.

6.2.2.   Filling gaps in the data

The available data on the Rhine and Meuse consist of several gaps. Some of the gaps consist of a few missing values and are scattered throughout the time series whereas other gaps are longer and consist of consecutive missing values for a few days at a time. Three different approaches are followed to fill in the missing data:

❑ *Linear interpolation*: when the number of consecutive missing values in the data is not more than five, the missing values are filled by linearly interpolating between the records preceding and following the gap.

❑ *Stage-discharge relationship*: at locations where separate discharge and water level records and a relatively stable stage-discharge relationship are available, such as Borgharen, the stage-discharge relationship is used to fill the gaps.

❑ *Neural network model*: when the gaps are large, interpolation within the time series is rather crude. Also, the stage-discharge approach might not be applicable if both discharge and water level are recorded or they are not correlated (as at Hagestein). In such circumstances, an ANN model that uses data upstream and downstream of the intended location is applied maintaining the proper lag times between the locations. The lag times are selected based on the average information travel time obtained from the AMI analysis. For example, the wider gaps in the data at Venlo are filled using data at Borgharen and Megen.

6.2.3.   Operation of control gates

Knowledge of the gate operations is important since it alters the natural flow pattern in the river. It poses a considerable uncertainty in operating forecast models since the way in which the gates are operated might not be known within the forecast period. Obviously, the level of uncertainty depends on the extent to which the gates affect the flow.

On the Rhine River, there are three gates located on the Lower-Rhine branch at Driel, Amerongen and Hagestein (Figure 6.2). The gate at Driel is the one that regulates the flow distribution along the different branches of the river. The regulation is mainly done when low discharges (water levels) are observed at Lobith. The operation rules are as follows:

❑ The gate at Driel is regulated when the flow at Lobith is less than 1500 m^3/s.

❑ Driel is completely open when the flow at Lobith is greater than 2320 m^3/s.

❑ All gates on the Lower-Rhine are raised when the flow at Lobith is greater than 3520 m^3/s or when the flow at Driel is greater than 638 m^3/s.

Data on the intended target flow distribution in the Rhine branches are available. Yet the data have to be checked with the actual observed distribution. It is important to analyse the relationship between the water level at Lobith and the gate operation at Driel. This helps to check the consistency of the gate operations. Figure 6.8 (a) & (b) show scatter diagrams of the water level at Lobith versus the difference in water levels upstream and downstream of Driel for the periods between 1996 to 1997 and 1998 to 1999 respectively. The difference in

water level at Driel is used as an indirect indicator of the gate operation since such data was
not available.

A visual observation of the scatter diagrams shows that there is a more or less consistent
pattern in the gate operation during the two periods. It also reveals that conditions at Driel are
fairly predictable since the operation pattern shows four distinct zones that can be
approximated, for example, with piece-wise linear functions. This is useful since flow
forecasting at Tiel and Hagestein beyond a horizon of 6 h might need the forecasting of the
status of the gate at Driel. The pattern in Figure 6.8(b) appears to have fewer outlying points
than the one in Figure 6.8(a), suggesting that gate operation at Driel is more consistent in the
later years. There is no evidence if the deviant operations in Figure 6.8(a) resulted from
human error or other operating criteria.

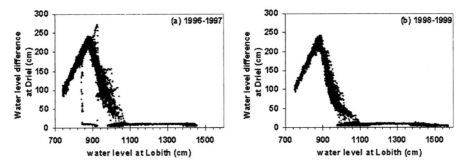

Figure 6.8. Gate operation at Driel versus water level at Lobith

On the Meuse River, all seven gates between Borgharen and Lith are raised for discharges
greater than 1250 m^3/s. There are no measurements available to take into consideration the
effect of the gates on flow forecasts.

6.2.4.   Stage-discharge relationship

For some of the measuring locations, both water level and flow records are available. In some
cases, the flows are measured directly. In others, the flows are mapped from water level
records using rating curves. Hysteresis and bed movement can affect the relationship between
stage and discharge. Hysteresis is caused by the attenuation of the hydrograph leading to
different wave speeds during the rising and falling limb: for the same river stage, the flood
wave moves faster during the rising limb of the hydrograph than during the recession. To
analyse the presence and extent of this effect, the data at Venlo on the Meuse River is
considered since both discharge and water level are measured separately.

Figure 6.9 shows a scatter diagram of flow and water level at Venlo. It can be observed that
there is no one-to-one relationship between the water level and the discharge. The presence of
loops particularly corresponding to water levels in excess of 1500 cm, suggests the effect of
hysteresis. The figure shows that the use of a one-to-one stage-discharge relationship can
result in errors in discharge of up to 100 m^3/s. This affects the accuracy of flow forecasts
made based on upstream water level and a rating curve. Lobith is an example where water
level measurements are used. Figure 6.10 shows the rating curve used at Lobith to convert
water levels to discharges. Since Lobith involves much higher discharges than Venlo, there
could be even wider loops resulting in larger errors in forecasted discharges downstream.

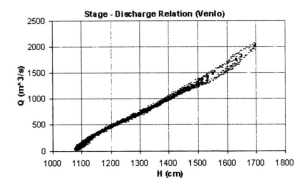

Figure 6.9. Stage-discharge relationship of the Meuse at Venlo

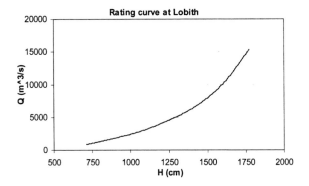

Figure 6.10. Rating curve of the Rhine at Lobith

Bed movement is also an important factor in the stability of the stage discharge relationships. Reports indicate that there are gradual changes in the bed levels of these rivers. For instance, the bed of the Waal has lowered by an average of 2 cm per year due to erosion and mining activities (RIZA, 1993). The effect of long-term averages might not be so significant compared to bed changes as a result of flood discharges. It is therefore necessary to evaluate forecast models with time.

6.2.5.   Tidal influence

The flow data at Hagestein and Tiel were analysed to check the presence of tidal effects from the North Sea. Three different methods were applied: the AMI analysis, evaluation of the accuracy of a naïve forecast model and Fourier spectral analysis.

The AMI analysis was carried out on hourly discharges at Hagestein. This helped to observe the rate of information deterioration with lag-time. Figure 6.11 shows the AMI for the first 72 hours. The graph shows that, along with an underlying trend of loss of information with time, there is a peculiar periodic peak with a period of nearly 12 and 25 h. The precise period cannot be computed with AMI analysis.

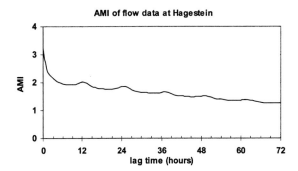

Figure 6.11. AMI analysis of hourly flow data at Hagestein

A naïve-forecast model, one that assumes that the forecast variable does not change after the last time it was observed, was applied at Hagestein. The accuracy of this model was tested up to a horizon of 72 hours. Figure 6.12 shows the mean absolute error between forecasted and observed flow versus forecast horizon. The graph shows that there is a periodic fall in the mean absolute error every 12 and 25 h.

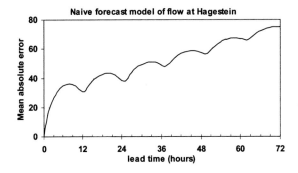

Figure 6.12. Accuracy of naïve forecast model at Hagestein versus forecast horizon

A Fourier spectral analysis on the time series of hourly difference in discharge at Hagestein showed that the data has a periodic component with a period of 12.42 h. In the context of this data, the primary physical process that can introduce such periodicity is the effect of tidal waves from the North Sea. The obvious implication is that, at Hagestein, tidal influence exists and has to be taken into account in some way.

Similar analyses were conducted on the flow data at Tiel, which showed no visible tidal presence. The analyses could not be repeated at Lith since the available flow data is not adequate.

6.3.   Model choices

The first modelling option considered is the use of a data-driven modelling approach, in particular, artificial neural network modelling. For this particular problem, ANN modelling is considered as a viable choice since it mainly exploits the relationship between upstream and downstream data. Such an approach does not need physical parameters of the river system even though such knowledge is an asset in the modelling process. It is particularly suitable for the Rhine branches since there is uncertainty in the flow distribution between the branches. However the input-output relationship has to be studied first. Here, the results of the AMI analysis carried out in §6.2.1 are used to select the appropriate inputs and the corresponding lag-times for each forecast location.

The second option considered is the use of a physically based routing model. Due to the limited scope of the project, the use of a 2-D model is ruled out. Since both river systems are protected with dikes and do not behave as natural meandering rivers, the 1-D assumption is expected to work reasonably well to forecast average flows over a cross-section (see Cunge et al. 1980). Again, the implementation of a 1-D model based on the full de Saint Venant equation would need both upstream and downstream boundary conditions, in addition to the initial conditions. Therefore, the same model used for the case study on the River Wye (§5.11) is considered here.

The model equation is used in its conservative form (equation (5.9)) to ensure mass conservation, particularly when c is taken not as a constant but rather as a function of the flow. Since the equation basically works with all practical Froude numbers and needs only upstream boundary and initial conditions, it is an ideal choice for a forecast model. The parameters c and d are functions of the flow and peculiar to each river reach. It has been shown by Price (1976) that the model can be applied assuming the whole river reach as a number of sub-reaches having similar interaction with the flow. Such a model has been applied on the Rivers Wye, Stour and Soar in the UK (Price 1973, Price 1985). An important issue in the application of the model in forecast mode is the upstream boundary data. The model needs upstream boundary data for the forecast period. In this study, the forecast horizon is limited to the average travel time from upstream to downstream. Thereafter, the upstream boundary data can be assumed to be constant and extended for the forecast horizon (the naïve model assumption) without compromising the forecast accuracy.

6.4.   Neural network routing model

The neural network routing model is based on a direct mapping of upstream conditions to downstream flows. Development of an ANN routing model involves three general steps:

Step 1: Selection of the appropriate set of input data for each forecast location depending on the available data and its relationship to the intended target output,

Step 2: Selection of appropriate lag time (information propagation time) between observation points, and

Step 3: Training and validation of the ANN model based on historical data.

The model development process implicitly includes insuring the availability of the input data to meet the required forecast horizon.

6.4.1.   Neural network forecasting on the Rhine

In order to forecast flows at Hagestein and Tiel, data at Lobith and Driel-downstream are considered. The AMI showed that average travel time from Lobith to both Tiel and Hagestein is 13 h, and from Driel to Tiel is 6 h. This travel time between Driel and Tiel does not imply that there are flows from Driel to Tiel, which are not located in the same branch of the Rhine. Nevertheless, it shows that information for an event observed at Tiel is related to that observed at Driel 6 h earlier.

Meeting the required forecast horizon of 24 h is an important factor in determining how many ANN models would be needed and what their input data would be. From the AMI analysis, the data observed at Driel allow only a 6 h lead-time whereas the data at Lobith allow 13 h. These lead-times can be increased but the ANN models would be of sub-optimal accuracy. To complete a 24 h forecast horizon at Tiel, a three-period forecast scheme is considered:

Period 1 (0-6 h): Using observed water levels both at Lobith and Driel.

Period 2 (6-12 h): Using observed water levels at Lobith and forecasted water levels at Driel. This needs a separate model to make forecasts at Driel based on water levels at Lobith. Another alternative is to use only the data at Lobith and assume that the ANN learns the effect of the gate on the flow at Tiel. This assumption is justified by the fact that the gate operation is dependent on the water level at Lobith.

Period 3 (12-24 h): Using forecasted water levels both at Lobith and Driel. This needs 12 h forecasts at Lobith and Driel. The other alternative is to use data from measuring locations in Germany.

The three forecast periods are expected to have a progressively reducing accuracy. However within each period, the order of accuracy should not depend on the actual horizon. Several tests were conducted using ANN models to forecast flows at Hagestein and Tiel, which led to the following observations:

❑ From period 1 (0-6 h) to period 2 (6-12 h) the mean absolute error (MAE) of the forecasts at Tiel increases from 30 m^3/s to 50 m^3/s. For Hagestein, there is only a slight increase from 33 m^3/s to 35 m^3/s.

❑ More tests showed that there is not much difference in forecast accuracy between using the water level at Lobith directly or after converting it to flow using the rating curve.

❑ Forecasting the water level at Tiel and converting it to flow using the rating curve shows the same order of accuracy as forecasting the flow directly.

6.4.2.   Neural network forecasting on the Meuse

For the Meuse River, the input data for an ANN forecast model could be the data at Borgharen or Venlo. The problem is, since there is no adequate flow data at Lith, the neural network cannot be developed by targeting the flow at Lith directly. To overcome this problem, the network is trained using the flow at Megen as a target. This is based on the notion that flow forecasts at Lith will be a time-lagged version of the hourly flow forecasts made at Megen. Assuming the same information travel speed in the reaches Venlo-Megen and Megen-Lith, the 15 km distance between Megen and Lith has a travel time of 2 h.

Since observed data at Borgharen enable flow forecasts at Lith 24 h in advance, there is no need to use forecasted flows at Borgharen. Therefore, the forecast at Lith is based on a two-period scheme:

Period 1 (0-8 h): Since the average travel time between Venlo and Megen is 6 h, forecasts at Lith can be made 8 h in advance using the observed flow data at Venlo. The mean absolute error of the flow forecasts at Megen for this period is 45 m^3/s. Using the data at both Borgharen and Venlo did not show any improvement.

Period 2 (8-24 h): Since the average information travel time between Borgharen to Megen is 22 h, implying that it is 24 h to Lith, a forecast horizon of 8-24 h can be covered with an ANN model using observed data at Borgharen. For this period the mean absolute error of the forecasts at Megen is 67 m^3/s.

6.5.   Physically based routing model

6.5.1.   The Rhine model

The physically based routing model is applied to the Rhine branches to route the flow from upstream to downstream. The model of the Rhine is perhaps the most complicated one for at least three reasons:

❑ The presence of a tidal influence at Hagestein makes forecasting flows at Hagestein based only on upstream flow conditions complicated.

❑ The flow bifurcation at Pannerden and IJssel junctions also complicates the flow forecast.

❑ The presence of the three gates at Driel, Amerongen and Hagestein on the Lower-Rhine influences the flow distribution, particularly at the IJssel junction. This human intervention is less predictable by any model.

In addition to the above reasons, there is also the effect of assuming a one-to-one stage-discharge relationship at Lobith that does not take the loop rating effect into account (see Figure 6.10). The approaches used to address these problems are discussed in the following sections.

Schematisation of the Rhine

The physically based model needs initial conditions and upstream boundary data. For the forecasts at Hagestein and Tiel, the water level at Lobith is used as the upstream condition. Since the information travel time from Lobith to Hagestein and Tiel is 13 h, forecasted water levels at Lobith must also be available in order to complete a 24 h forecast horizon at Hagestein and Tiel. Figure 6.13 shows the schematic representation and the number of space grids (JJ) selected for each reach on the Rhine branches. The IJssel branch is not included in the model since there is no forecast location on that reach. The space grid steps (Δx) are determined on the basis of information travel times corresponding to a time step (Δt) of 1 h to ensure the average Courant number ($c\Delta t/\Delta x$) close to unity.

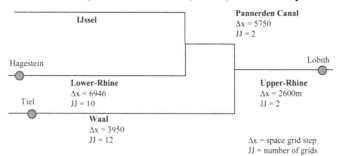

Figure 6.13. Schematisation of the Rhine branches

According to this model schematisation, the Upper-Rhine reach is computed first. Then a flow distribution scheme is used to determine the share of the flow going to the Waal and to the Pannerden Canal. The Waal branch is routed next, which completes the forecasts at Tiel. After routing the flow in the Pannerden Canal, a flow distribution scheme is used once again to determine the share of the flow to the Lower-Rhine and IJssel branches. Routing the flow along the Lower-Rhine completes the forecasts at Hagestein. At first a flow distribution table generated by a physically based model (SOBEK) was used as a look-up table both at the Pannerden and IJssel junctions.

Forecasting the flow at Tiel

The model was tested using part of the available historical data. The observed and modelled flows at Tiel for the year 1999 are shown in Figure 6.14. The results show that the model tends to consistently overestimate the low flows at Tiel. In order to confirm this, the model errors at Tiel are plotted against flows at Lobith in the scatter diagram shown in Figure 6.15.

The figure shows some peculiar features corresponding to different ranges of the flow at Lobith. There are clearly patterns between the errors and the flows at Lobith.

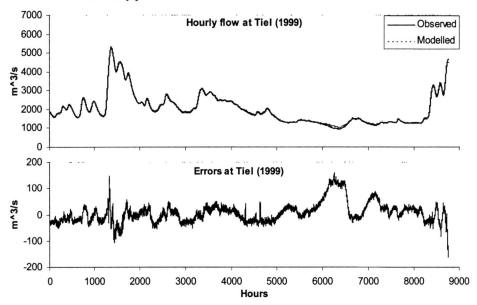

Figure 6.14. Physically based model forecasts at Tiel

To solve this problem, two alternatives are explored. The first alternative is to complement the model with a corrective scheme such as a piece-wise linear or ANN model that corrects the forecasts on the basis of water levels or flows at Lobith. The second alternative is to adjust the flow distribution at the Pannerden junction. The second alternative is preferred since the flow distribution at the Pannerden junction can actually be determined.

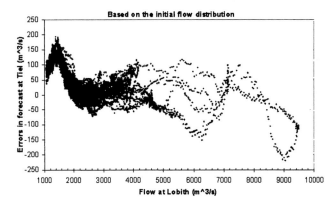

Figure 6.15. Forecast error at Tiel versus flow at Lobith

To determine the actual flow distribution, it is necessary to know the flow in the Waal corresponding to the flow in the Upper-Rhine. This is accomplished using a 'reverse' routing procedure. The reverse routing procedure is simply a rearranged form of the usual 'forward' routing procedure shown in equation (5.11) in such a way that the downstream boundary condition is used to determine the flow at a location upstream of the reach. Using this reverse routing, the flow at Tiel is used as a boundary condition and propagated back to the upstream end of the Waal branch. Similarly, using forward routing, the flow at Lobith is propagated to the downstream end of the Upper-Rhine branch. The scatter diagram shown in Figure 6.16 shows the flows in the Waal versus the flows in the Upper-Rhine, both close to the junction. The flow distribution pattern initially used is also shown.

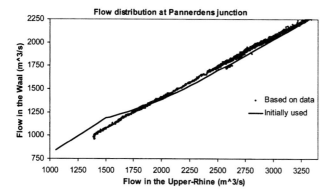

Figure 6.16. Flow distribution at Pannerden junction

From the graphs in Figure 6.16 it can be deduced that

❑ The flow distribution at Pannerden junction shows a more or less linear pattern and is not much influenced by the gates in the Lower-Rhine branch, or at least the effects have been overestimated.

❑ The flow distribution initially used has a tendency to overestimate the flows in the Waal for flows in the Upper-Rhine less than 1900 m^3/s, and to underestimate the flows in the Waal for flows in the Upper-Rhine more than 1900 m^3/s.

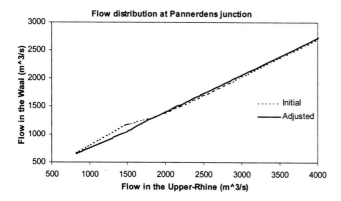

Figure 6.17. Flow distribution at Pannerden junction (initial and adjusted)

Therefore, it is necessary to adjust the flow distribution scheme at this junction in accordance with the observed data. The scatter diagram in Figure 6.16 is used to construct a new look-up table to determine the flow distribution at Pannerden junction. A piece-wise linear approach is used to develop the new distribution shown in Figure 6.17.

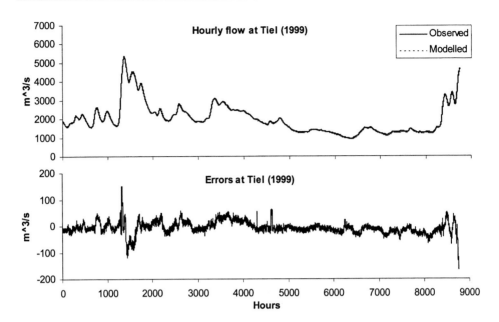

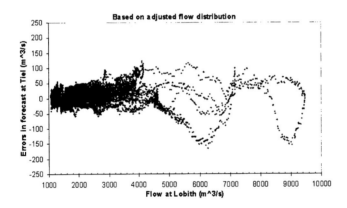

Figure 6.18. Physically based model forecasts at Tiel (using adjusted flow distribution)

Figure 6.19. Forecast error at Tiel versus flow at Lobith (using adjusted flow distribution)

Forecasts at Tiel are redone using the adjusted flow distribution at Pannerden junction. The resulting flow time series at Tiel for the year 1999 is shown in Figure 6.18 along with the forecast errors. The new results show that the forecast accuracy is improved dramatically, the mean absolute error falling from 29 m³/s to 16 m³/s. Also, as shown in Figure 6.19, the errors no longer demonstrate any obvious patterns with the flow at Lobith. However, there are loops corresponding to high discharges at Lobith, which could be the effect of hysteresis.

Forecasting the flow at Hagestein

The forecasted flows at Hagestein for the year 1999 are shown in Figure 6.20. The forecasts show some of the expected problems. The presence of negative flow records and the periodic surges in the data compared to the relatively smooth line from the model indicate the tidal effect. The selected model cannot incorporate tidal effects, but neither can any other model without the use of a downstream boundary forcing of some sort. From the time series of forecast errors at Hagestein in Figure 6.20 it can be seen that the errors at Hagestein have higher magnitudes than those at Tiel.

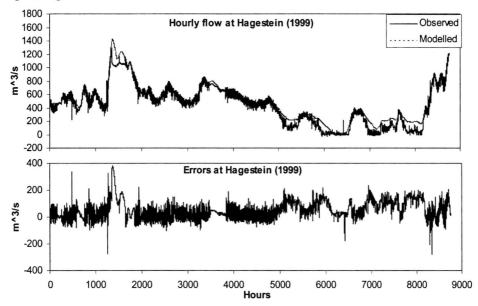

Figure 6.20. Physically based model forecasts at Hagestein

Adjustment of the flow distribution at IJssel junction is not considered at this stage since, unlike the data at Tiel, the data at Hagestein is affected by tidal components. The other alternative is to use flow data on the IJssel branch, which was not available for this study. Figure 6.21 shows flow at Hagestein versus flow at Lobith lagged by 12 h. It can be seen that the relationship between flow data at Lobith and Hagestein is more complicated than the relationship between the flow data at Lobith and Tiel. The effect of the control gates across the Lower-Rhine is clearly visible, particularly corresponding to low flows at Lobith.

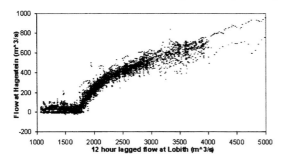

Figure 6.21. Flow at Hagestein versus 12 h lagged flow at Lobith

The potential sources of forecast errors at Hagestein are the tide and the operation pattern of the gates. The operation of the gates affects the accuracy of the forecast at Hagestein not only by altering the flow distribution at the IJssel junction but also by introducing attenuation in the flow. As long as the gate operation in the Lower-Rhine is consistent, the routing model does not have to make any direct provisions. Even if it did, it would not be possible to take the operation of the gates into account in forecast mode other than by incorporating it into the flow distribution pattern. The accuracy of the forecast model at Hagestein is therefore subject to the condition that gates on the Lower-Rhine branch are operated with a consistent operational policy.

Correction for tidal effects

It has already been established in §6.2.5 that there is tidal influence at Hagestein, which has to be accounted for. Two alternatives are considered. The first alternative is to add a periodic component to the forecasted flows based on a harmonic analysis of the errors. The danger with this approach is that even tiny errors in the period of the components could cause a phase error that accumulates with time and create more discrepancy. The second alternative, which is applied here, is the use of a complementary model to forecast future errors from past errors.

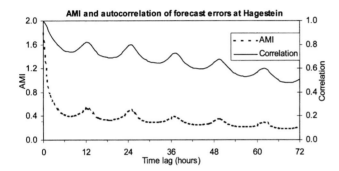

Figure 6.22. AMI and autocorrelation of forecast errors at Hagestein

At first, the AMI and autocorrelation of forecast errors are computed at Hagestein up to a lag-time of 72 h as shown in Figure 6.22. The figure shows that the model errors share maximum levels of information at lag times of 12 and 25 h. It has to be noted that the AMI approach can resolve periodicity only to the time step of the data used. Therefore it is not possible to find out the exact period of the components. However, the exact period of the waves is not necessary in the approach followed to update the forecasts. It is possible to forecast the errors at Hagestein based on the errors 12 h or 25 h earlier. Since the forecast horizon is set to 24 h, a correction scheme that uses the errors 25 h earlier is adopted. If 25 h earlier errors are needed as input, this in turn implies that the error-forecast model needs latest observed flow data at Hagestein (for the previous 25 h) in order to compute the errors and use them in forecasting the future errors.

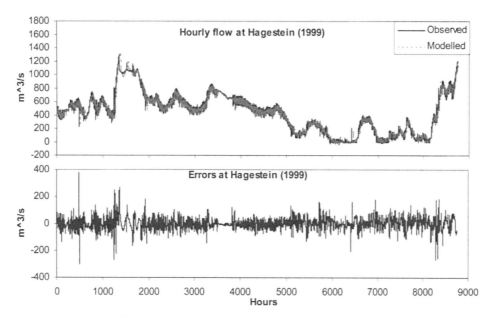

Figure 6.23. Physically based model forecasts at Hagestein with correction for tidal effects

Since the linear correlation of the errors separated by 25 h is high, a linear correction model was developed and applied on the forecasts made by the physically based model. Figure 6.23 shows the observed and forecasted flow at Hagestein after incorporating the error correction scheme. The forecast accuracy is improved substantially with the mean absolute error falling from 60 m^3/s to 29 m^3/s. Figure 6.23 also shows that because of the correction for tidal effects, the error magnitudes are much lower compared to those shown in Figure 6.20.

6.5.2.   The Meuse model

The model for the Meuse is intended to make 24 h in advance forecasts at Lith. Compared to the Rhine, the Meuse River is easier to model since it is one long reach between Borgharen and Lith without any branches. Nevertheless, it has its own problems, which are:

❑ Parts of the river reach have undergone changes since 1995 (WL | Delft Hydraulics, 2001). The extent to which these activities alter the property of the flow has to be studied. The data used to develop and verify the model were recorded between 1997 and 2001, which are rather recent. However, the model performance has to be checked regularly.

❑ At Lith, which is the target location, water level data is available at 10-minute intervals, but the flow records are at daily intervals. This implies that it is not possible to verify the model with the data at Lith. Therefore the model has to be developed targeting the flow at Megen. The model domain is then extended downstream to Lith by 15 km assuming the same flow conditions as the preceding reach.

❑ The effect of the seven gates across the river on the flow, especially during low flows, is not easy to incorporate in a forecast model.

Schematisation of the Meuse

The part of the Meuse between Borgharen and Lith is schematised for the physically based routing model as shown in Figure 6.24. The model domain is divided into three reaches: Borgharen-Venlo, Venlo-Megen and Megen-Lith. Once again, the space grids are defined based on the length of the reach and the average information travel time along the reach obtained from the AMI analysis.

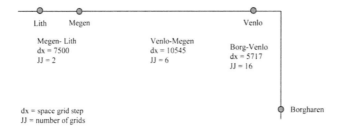

Figure 6.24. Schematisation of the Meuse

The average travel time between Venlo and Megen is 6 h. Assuming the same travel speed between Megen and Lith gives a travel time of 2 h. This means, using the data at Venlo, it is possible to forecast flow at Lith up to 8 h in advance. To complete a 24 h forecast horizon, the flow at Borgharen has to be used. Since the average travel time between Borgharen and Venlo is 16 h, it is possible to make 24 h in advance forecast at Lith using only the data on the Dutch part of the Meuse.

The lateral inflow is found to be very important particularly in the reach Borgharen-Venlo. Mass balance analysis showed that the average lateral inflow along the reach is 62 m³/s, which is quite considerable. The three reaches on the Meuse are considered step by step in the following sections.

Borgharen - Venlo

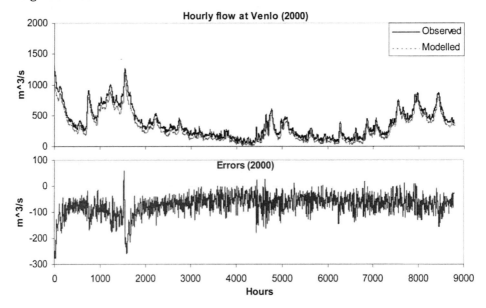

Figure 6.25. Physically based model forecasts (Borgharen-Venlo)

The Borgharen-Venlo reach is considered first. The convection speed was calculated using the average information travel time obtained using the AMI analysis. The attenuation parameter was determined by automatic calibration within bounds. Figure 6.25 shows the observed and modelled flows at Venlo. The graph shows that the model consistently underestimates the flow at Venlo thus the errors show a considerable negative bias. The mean absolute error is 72 m³/s, which is quite high compared to that of the Rhine model. It is therefore necessary to investigate the actual cause of such a high discrepancy between the observed and forecasted flows at Venlo. It could be due to the model parameters, the lateral inflow or the effect of the gates.

Monte Carlo Simulation at Venlo

A Monte Carlo simulation was conducted to analyse the effect of the model parameters on the forecast accuracy. The parameters considered are the convection speed c and the attenuation parameter d. The convection speed was generated using a Gaussian distribution with a mean of 1.86 m/s (obtained using the average travel time) and a standard deviation of 0.4 m/s. The attenuation parameter was generated using a mean value 6420 m (obtained by calibration) and a standard deviation of half the mean. A hundred pairs of parameters were generated and model simulations were carried out between Borgharen and Venlo. Figure 6.26 shows the time series of the standard deviation of the flows obtained from the 100 simulations. This time series is basically the same as the time series of the standard deviation of the forecast errors. The time series of model errors obtained using the mean values of c and d is also shown in the same graph.

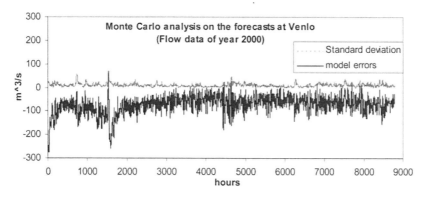

Figure 6.26. Monte Carlo analysis on physically based model forecasts at Venlo

The following observations are made from the graph:

□ The variation due to the parameters is far less than the actual simulation errors, which suggests that the model parameters are not the main cause of the discrepancy between observed and forecasted flows at Venlo.

□ The model errors are consistently negative showing that the model underestimates the flow at Venlo, which is an indication that there is a considerable lateral inflow between Borgharen and Venlo.

□ The magnitude of the error increases with the discharge at Venlo. This indicates that the lateral inflow tends to increase when the discharge in the river is high and vice versa, which is not unreasonable.

Analysis of heteroscedacity of the errors at Venlo

In addition to the Monte Carlo simulation, further analysis of the errors of the physically based model forecasts at Venlo is considered. The error time series of model forecasts at Venlo (Figure 6.25) shows a peculiar pattern in that the variation of the errors increases with the observed discharge. Such a pattern is statistically known as heteroscedacity. Heteroscedacity is characterised by the presence of a variation in the variance of the errors in relation to some independent variable. When the variance is more or less constant, the error time series is known to be homoscedatic.

In some cases, the actual error time series might not show a clear pattern with respect to some other time series by the use of correlation and average mutual information; whereas its variance might. One way to detect the presence of heteroscedacity in the error time series is to compute the moving variance of the errors and to observe its variation with respect to time as well as the observed discharge.

Consequently, the moving variance of the errors was computed for the data of the year 2000 using moving windows of 20, 30 and 50 time steps (hours). The results are shown in Figure 6.27 along with the time series of the observed discharge. The figure shows that the variances tend to increase with discharge. There is also a considerable variation of variances with time. The ratio between the maximum and average variances corresponding to windows of 20, 30 and 50 time steps on the year 2000 correspond to 15.27, 15.25 and 19.60 respectively. These two conditions indicate the presence of heteroscedacity in the error time series.

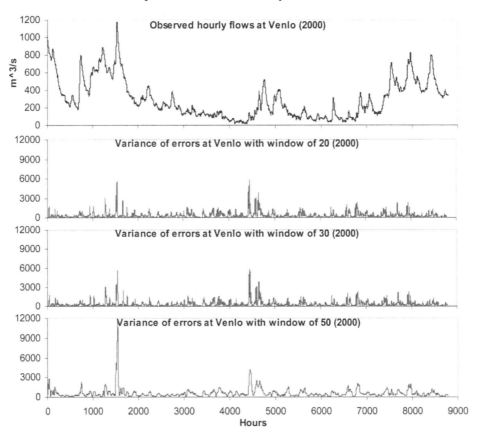

Figure 6.27. Moving variance of model errors at Venlo

There are no established guidelines to overcome heteroscedacity in the literature. One approach used is to rescale the errors with respect to the independent variable with which the error is heteroscedatic before developing any model. This is essentially a weighting process that takes care of the variation with respect to this independent variable. The modelling will be done using the scaled errors and the errors are later scaled back in order to get their actual values, which will be used to update the model forecasts. This procedure was applied by Mohamed Nanseer (2003) on a problem of uncertainty analysis of the Hupselsebeek, The Netherlands. The procedure follows scaling of the errors in such a way that the error time series will be distributed as uniformly as possible with respect to the observed discharge.

In this approach, the model residuals $\{R_n\}_{n=1}^{N}$ and the variance, v_n, of the nth residual are related as:

$$v_n = \sigma_n^2 = aQo_n + b \tag{6.1}$$

where Qo_n is the discharge corresponding to the nth residual and a and b are parameters to be identified by minimising a log-likelihood function L that can be described as:

$$L = \tfrac{1}{2}\sum_{n=1}^{N} \ln(v_n) + \tfrac{1}{2}\sum_{n=1}^{N} R_n^2 \frac{1}{v_n} \tag{6.2}$$

so that L is actually a function of a and b, and N is the total number of residual ordinates.

The nth ordinate of the new (scaled) residual series is defined as $\dfrac{R_n}{\sqrt{v_n}}$.

The parameters a and b determined by minimising the function L will later be used to compute the scaling factors v_n in order to rescale back the errors to their actual values.

For the Meuse data, the minimisation of L results in the values of parameters $a = 1.7255$ and $b = 70.5$. These values are used to rescale the errors. The scaled errors for the year 2000 data are shown in Figure 6.28 along with the corresponding variance computed with a window of 50. The ratio of the maximum to the average variances for the year 2000 data is 8.77 for the scaled errors (compared to 19.60 without scaling). This indicates that the scaling process actually reduces the variation of the variances, which also proves the presence of some degree of heteroscedacity with respect to the observed discharges.

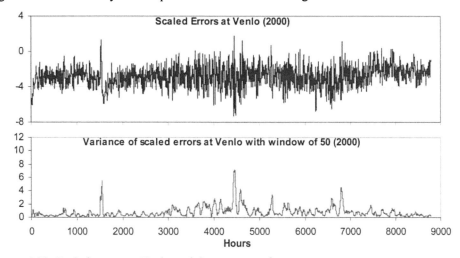

Figure 6.28. Scaled errors at Venlo and the corresponding moving variance

Complementary modelling at Venlo

Since the lateral inflow appears to show a pattern related to the discharge in the river, the use of a constant lateral inflow term based on the mass balance causes positive errors for low flows and negative errors for high flows. Due to this reason the complementary modelling procedure is considered at Venlo. An AMI analysis was carried out between the forecast error at Venlo and some state variables. The analysis showed that the error shares maximum information with the flow at Borgharen and its first time derivative, in both cases the lag time being 19 h. It also showed that the error is related to the magnitude of the flow at Venlo at the same time step.

Two different procedures of applying complementary modelling are considered. The first procedure is to attempt to forecast the errors of the physically based model directly with the use of an ANN model. The neural network is trained to predict the forecast errors of the physically based model at Venlo using as input the discharge at Borgharen, its first time derivative and the initial physically based model forecasts at Venlo. The graphs in Figure 6.29 show the combined performance of the physically based model and this complementary ANN model. The mean absolute error dropped from 72 to 28 m³/s. Improving forecasts at Venlo means improving forecasts at Megen and Lith as well.

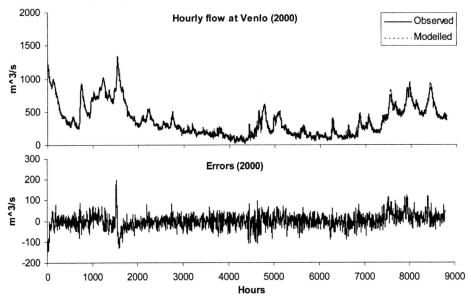

Figure 6.29. Physically based model (Borgharen-Venlo) and ANN update at Venlo

The other procedure followed in the development a complementary model is the use of an ANN model that targets the scaled errors instead of the actual errors.

The procedure is as follows:

1. Compute the scaling parameters (a and b) for the error time series using the data that will be used for training.

2. Scale the errors of the physically based model corresponding to both the training and verification data sets.

3. Train an ANN error prediction model using the training data set. The target output of the ANN is the scaled error.

4. Check the performance of the ANN on the scaled errors of the verification data set.

5. Scale back the ANN predicted errors of the verification data set using the scaling factors computed on the basis of the discharges predicted by the physically based model. The observed discharges cannot be used in the scaling back process since these data are not available in forecast-mode.

6. Apply the scaled-back errors to update the preliminary forecasts of the physically based model and compare the resulting updated discharge forecasts with the observed discharges at Venlo.

This procedure is applied and the forecasted discharges are consequently updated. The same input data is used for the complementary ANN model as in the one applied without rescaling the errors. The observed and updated discharges at Venlo and the corresponding residuals for the year 2000 are shown in Figure 6.30. The mean absolute error corresponding to the verification part of the data is 27 m³/s. There is also a reduction in one of the positive peak errors (between 1000 and 2000 hours on the time axis) compared to the errors shown in Figure 6.29. The overall improvement in mean absolute error is not high compared to the direct application of a complementary ANN model without rescaling the errors. However, the fact that there is a reduced variation in the variance of the scaled errors and the slight improvement in accuracy gives a good reason to pursue the technique for further research with a number of case studies.

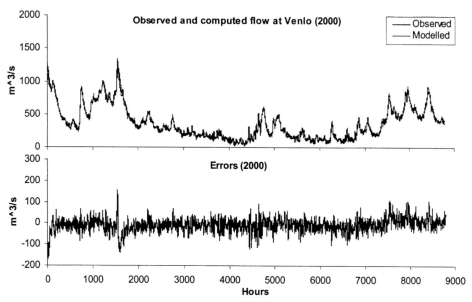

Figure 6.30. Physically based model forecasts at Venlo updated with ANN model of scaled errors

Venlo - Megen

The performance of the physically based model in the reach between Venlo and Megen was much better than in the preceding reach. This is possibly because the reach is shorter and does not have considerable later inflow. With the use of the observed hourly flows at Venlo, the mean absolute error between forecasts and observations at Megen is 30 m³/s. In addition, the errors do not show any visible pattern and a complementary model is not considered. The graphs in Figure 6.31 show the observed and modelled hourly flows at Megen for the year 2000.

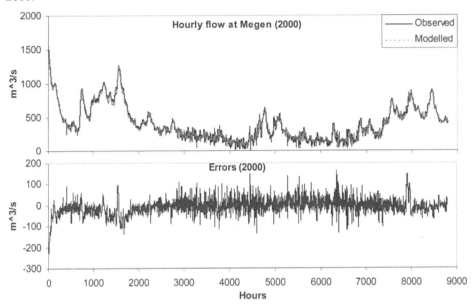

Figure 6.31. Physically based model forecasts (Venlo-Megen)

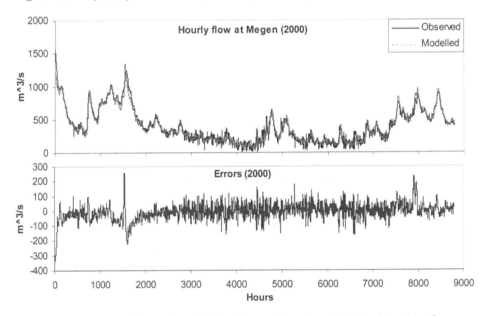

Figure 6.32. Physically based model (Borgharen-Megen) and ANN update at Venlo

According to the AMI analysis, forecasts at Megen using the data at Venlo can only be made up to a horizon of about 6 h. For the remaining 16 h (18 h to Lith), the forecasts are dependent on observed flows at Borgharen. The forecasts will obviously be less accurate than those obtained using the data at Venlo. However, if ANN updating is done at Venlo, the results are much better showing a mean absolute error of 45 m^3/s. Figure 6.32 shows the observed and forecasted flows at Megen using the physically based model along with updating at Venlo by the use of the complementary ANN model.

Forecasting the flow at Lith

The ultimate purpose of the Meuse model is to make 24 h in advance flow forecasts at Lith. Forecasts at Lith are made by extending the model domain by the reach from Megen to Lith assuming the same parameters as in the preceding reach. Since there is no hourly flow data at Lith, the accuracy of forecasts at Lith cannot be validated. However, the Megen-Lith reach is relatively short and it is reasonable to assume that the forecast accuracy at Lith has the same order of magnitude as that of Megen.

6.5.3.   Discontinuity tests

The flows forecasted by the models developed under this study are intended for use as boundary conditions for another model that covers the delta and part of the North Sea. Since jumps in boundary data are known to cause computational instability in hydrodynamic models, it is necessary to check for the presence of jumps as a result of the forecasts made by the forecast models. There are possible causes for jumps in both the Rhine and Meuse models. The intention here is to check whether the jumps introduced numerically are higher than naturally occurring variations in hourly discharge at the forecast locations Tiel, Hagestein and Lith. The assumption here is that if the numerically induced jumps at forecast locations are less than naturally observed hourly differences in discharge, there is no threat of these jumps on the computational stability of the delta model.

For the Rhine model, a possible cause for a jump is the use of two different sources of upstream boundary data. The boundary data at Lobith comes from latest observations for 0-12 h horizon. For the remaining 12-24 h, the water level at Lobith is obtained from a separate forecast model. If there is a large variation in the transition between the observed and forecasted boundary conditions at Lobith, the effect propagates as a jump downstream to Tiel and Hagestein.

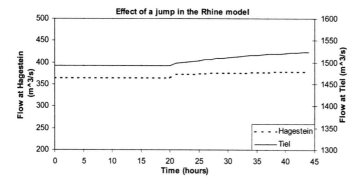

Figure 6.33. Effect of a 10 cm positive jump at Lobith on forecasts at Hagestein and Tiel

It was learned that the accuracy of water level forecast at Lobith falls within ±10 cm. Figure 6.33 shows how a positive jump of 10 cm, which is equivalent to a discharge of 58 m^3/s, introduced at Lobith translates to forecasted flows at Tiel and Hagestein. The results show that in both cases the corresponding change in flow spreads over 25 h period, and the maximum magnitude of the hourly variation in discharge is 10 m^3/s, which is much less than some of the hourly variations observed at Hagestein and Tiel, and is not expected to cause computational problems on the delta model.

For the Meuse model it is known that 24 h in advance forecasting at Lith can be accomplished using observed flows at Borgharen. However, to obtain more accurate forecasts for the first 8 h, the model uses the flow at Venlo as a boundary condition. One potential source of a jump is therefore the transition between the observed and the forecasted flows at Venlo. A positive jump of 400 m^3/s, which is a very high magnitude compared to the forecast errors expected at Venlo, is introduced into the Meuse model at Venlo and propagated to Megen and Lith.

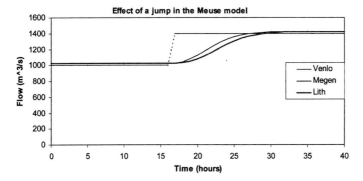

Figure 6.34. Effect of a 400 m^3/s positive jump at Venlo on forecasts at Megen and Lith

Figure 6.34 shows how forecasted flows at Megen and Lith are affected by a positive jump of 400 m^3/s introduced at Venlo. The results show that in both cases the corresponding change in flow spreads over a 15 h period, and the maximum hourly variation at Lith is less than 50 m^3/s, which again is much less than many hourly variations in the observed data, and is not expected to cause computational problems on the delta model

6.6.   Comparison of model choices

Accuracy

The accuracy of the forecasts at Tiel, Hagestein and Lith obtained from several model choices is shown in Table 6.1, Table 6.2 and Table 6.3 respectively. The mean absolute error (MAE) is used for the comparison. The tables also show the accuracy for each forecast location with different sets of input data. The choice of input data is an important factor since it affects the forecast horizon and because the amount of information available in each input is different. For instance, forecasts at Lith are more accurate when the input data is taken at Venlo rather than at Borgharen. However, if only the data at Venlo is used, the accuracy of the forecasts at Lith is expected to decrease for a forecast horizon beyond 8 h. The difference in accuracy between the physically based and ANN model forecasts is not so significant. Therefore, other aspects of the model choices have to be considered.

Table 6.1. Accuracy of forecasts at Tiel

Model choice	Input data	Horizon (h)	MAE (m³/s)
ANN model	H at Lobith	1–12	33
	H at Lobith and Driel	1–6	50
Physically based model	H and Q-H relationship at Lobith and flow distribution at junctions	1–12	29
Physically based model	H and Q-H relationship at Lobith and flow distribution at junctions	1–6	16

Table 6.2. Accuracy of forecasts at Hagestein

Model choice	Input data	Horizon (h)	MAE (m³/s)
ANN model	H at Lobith	1–12	33
	H at Lobith and Driel	1–6	35
Physically based model	H and Q-H relationship at Lobith and flow distribution at junctions	1–12	60
Physically based model and error correction scheme	H and Q-H relationship at Lobith, flow distribution at junctions, and past flows at Hagestein	1–12	29

Table 6.3. Accuracy of forecasts at Lith (errors are calculated at Megen)

Model choice	Input data	Horizon (h)	MAE (m³/s)
ANN model	Q at Venlo	1–8	45
	Q at Borgharen	9–24	67
	Q at Borgharen and Venlo	1–8	45
Physically based model	Q at Venlo	1–8	30
	Q at Borgharen	9–24	80
Physically based model and ANN correction at Venlo	Q at Borgharen	9–24	45

Input data

The input data used by the forecast models have to be available in forecast mode. For the ANN models, past upstream data are used to forecast future downstream flows on the basis of an established lead-time used during model training. The lead-time is set based on the information travel time from upstream to downstream. To run the physically based model, upstream boundary data has to be used at each time step for which forecasts are made. In this study, the last observed upstream data are assumed to be constant and extended till the end of the forecast period. The effect of the travel times is embedded in the convection speed. Changes at the upstream boundary will not show immediately at the downstream end. However, it is obvious that after a time approximately equal to the travel time, the effect will reach and affect the forecast accuracy. The physically based model also needs initial conditions at all grid points, which, in this case is assumed constant and equal to the flow at the upstream boundary at the beginning of the forecast. In order to avoid the effect of initial conditions on the forecast accuracy, the model has to be initialised some time back in the past.

For the ANN model, the upstream data can be water level or flow, whichever is used during training, since the model is assumed to learn the Q-H relationship as well. For the physically based model, only flow data can be used to forecast the downstream flow. If water level has to be used as a boundary condition, then it is necessary to use a Q-H relationship in the model.

Flow distribution

For the Rhine model, the flow distribution to the different branches is perhaps the most important aspect of the forecasting. Since the distribution largely depends on the flow in the Upper-Rhine, it can be assumed that the neural network model learns the distribution based on the water levels at Lobith. For the physically based model, there is no direct way to include the flow distribution in the model other than by providing an explicit distribution scheme. If the distribution pattern changes in the future for some reason, the physically based model can still work as long as it is provided with the new distribution scheme, whereas the neural network model has to be retrained after sufficient observations become available with the altered flow distribution.

Gate operations

The effect of the gate at Driel on the Lower-Rhine branch, which regulates the flow distribution mainly between the IJssel and Lower-Rhine branches, is an important factor. For the neural network model, the fact that the gate operation policy is based on the water level at Lobith takes care of its consequences. However, if the gate operation deviates from the one learned from the data, it will be a source of uncertainty, particularly for forecasts at Hagestein. The possibility of deviant operations has been observed from the data (see Figure 6.8(a)).

One possible way to include the status of the gate at Driel would be to use the difference in water levels between the upstream and downstream of the gate. Even though the performance of the ANN model using this data improves the forecast accuracy slightly, the advantage is limited to the first few hours of the forecast horizon since the average travel time is only 6 h.

Extended physical domain

For the Meuse River, the ANN model is trained with flow at Megen since there are insufficient data at Lith. The forecasted flows at Lith are then the time-lagged form of the forecasted flows at Megen, which is a procedure that does not account for the attenuation in the reach. For the physically based model, the appropriate procedure is to extend the model domain by the reach from Megen to Lith, in which case the forecasts at Lith are not just a time-lagged version of those at Megen. Instead the physics of the flow in the reach is also included. In the case of extensive maintenance on the river, the physically based model has an obvious advantage since it uses physical domain data and parameters with physical meaning, which makes it easier to update.

6.7. Conclusions and discussion

The models developed to make the forecasts face several sources of uncertainty. For the Meuse model, the ongoing construction on parts of the river may cause large deviations between forecasts and actual events. It is therefore important to review the forecast accuracy from time to time. The convection speed and attenuation parameter of the physically based model have to be updated if major changes are made to the river channel or if forecast discrepancies increase. For the model of the Rhine, an important factor for a reliable forecast is the flow distribution at the bifurcation points. The data on the Waal (at Tiel) do not show an observable sensitivity to the gate operations on the Lower-Rhine (particularly the one at Driel). Therefore the forecast accuracy is expected to remain as good as the one shown by the past records used in verifying the model. The forecast at Hagestein on the Lower-Rhine is most vulnerable to uncertainty associated with the gate operations. The gate at Driel affects the flow distribution, whereas all the gates alter the nature of the flow by introducing attenuation. It is impossible to incorporate fully the gate operations in forecast mode. As long as the operation policy remains consistent, the model is expected to perform well. If the gate operation policy changes, then it will be necessary to update the flow distribution scheme used by the model accordingly.

As in the case of the Pannerden junction, it is recommended that a similar procedure be followed to establish the flow distribution at IJssel junction to the IJssel and Lower-Rhine branches. This would need the use of flow data on the IJssel branch. Flow data on the Lower-Rhine cannot be used in the reverse routing procedure described in §6.5.1 because of the gate intervention and tidal influence.

Information theory-based principles played an important role in developing models for both rivers. AMI analysis helped to study the relationship between water level, discharge and gate openings observed at different locations. Using AMI measures, it was possible to understand how data measured at one location related to past records. This in turn helped to detect the presence of tidal influence at the measuring locations. It was also used to estimate the convection speed on the basis of the information propagation time between data measuring locations. By analysing the residual time series at Hagestein, it was possible to detect the presence of tidal influence from the North Sea and subsequently develop a complementary model that takes tidal effects into account. For the Meuse, AMI measures helped to determine the best related input data for the complementary ANN model that predicts the errors at Venlo.

The study showed that the complementary modelling approach is the best choice to make forecasts on both rivers. Using complementary modelling, physically based forecast models of the Rhine and Meuse have been able to incorporate processes that were initially ignored. For the Rhine model, the effect of the tide from the North Sea on the flow at Hagestein has been accounted for. For the Meuse model, the rather significant amount of lateral inflow between Borgharen and Venlo has been accounted for. In both cases, these processes were not included in the physically based model.

For the Meuse model, a scaling process is used to reduce the presence of heteroscedacity in the errors. The procedure helped to reduce the variation of the variances of errors with respect to time. The scaled errors were also used in the development of a complementary model. However, there was only a slight improvement in the accuracy when scaled errors were used compared to using the actual errors in the development of the complementary model. That could perhaps be because, in both cases, the discharge at Venlo is used as one of the inputs to the complementary ANN model. Nevertheless, the fact that the scaling procedure minimises the heteroscedatic behaviour of the errors suggests the importance of analysing the statistical behaviour of the errors in the development of complementary models for error prediction.

The data-driven modelling techniques used to develop the complementary models for the models of the two rivers are different. For the Meuse, an ANN model is used whereas for the Rhine, a linear model is used, which shows that, in principle, any data-driven modelling technique, which is appropriate for the type of the process and the available data, can be used to establish the complementary model. The other difference lies on the set-up of the complementary model. For the Meuse model, updating is done on an intermediate output of the primary model whereas for the Rhine model, updating is done on the final output of the primary model. This shows that complementary modelling can be applied at intermediate stages of the primary model as long as its importance in improving the forecast accuracy is anticipated.

CHAPTER 7. FORECASTING THE ACCURACY OF NUMERICAL SURGE FORECASTS ALONG THE DUTCH COAST

This chapter presents an extensive application of information theory-based principles and artificial intelligent modelling techniques to the problem of forecasting the accuracy of numerical surge forecasts along the Dutch coast. Unlike the other case studies which are done on rainfall-runoff and river models, this case study extends the dimension of the complementary modelling approach to a more complex model of the North Sea that involves meteorological surges and tidal waves. It presents a unique challenge in that direct error forecasting does not improve the forecast accuracy as much as it does in the other case studies. Rather the problem focuses on forecasting the accuracy of surge forecasts in the form of a linguistic description of the forecast accuracy and bias and confidence intervals using artificial neural network and geno-fuzzy rule-based modelling techniques. An attempt is also made to trace back error patterns to possible ways of improving the numerical model.

7.1. Introduction

This study is a part of the Rijkwaterstaat's Nautilus 10.10 project, the second phase of which is intended to develop a methodology of forecasting the accuracy of numerical surge forecasts made by the North Sea simulation models at specific sites along the Dutch coast, particularly at Hook of Holland.

The forecasts made by the North Sea simulation models are used at the seaward boundary of the Zeedelta model. This model covers the delta area of the Rhine and Meuse rivers and part of the North Sea adjacent to the Dutch coast. Its purpose is to forecast the currents and water levels as they affect navigation of shipping entering and leaving the port of Rotterdam. It also helps to forecast the saline intrusion that may affect drinking water intakes. The accuracy of the seaward boundary condition as generated by the North Sea simulation models is therefore important for the Zeedelta model.

The target is to generate a connectionist (or data-driven) forecast model that links the surge prediction accuracy at sites along the Dutch coast to the significantly related variables in the North Sea. It is therefore important that the input to this model be available in operational forecast mode.

Along with operational surge forecasts made by the oceanic model, the resulting data-driven model developed in this study can be used to forecast the accuracy of the expected surges. Having a forecast of the accuracy can help in determining the amount of uncertainty involved in any activity that involves the use of forecasted surges. In this study, three different measures of accuracy are used. These are:

❑ the deterministic value of the expected surge prediction error,

❑ the interval in which the surge can vary with a predefined confidence limit, and

❑ a linguistic description of the expected forecast accuracy.

7.1.1. The DCSM model

The Dutch Continental Shelf Model (DCSM) is a computational model used by the National Institute for Coastal and Marine Management (RIKZ). This 2-D hydrodynamic model computes water levels, currents and particle transport in open water. DCSM considers a study area that can be characterized by curviform shaped, quadrangular grid of computation points. Grid sizes for coastal areas vary from 30 m to 16 km. DCSM uses polar coordinates. The grid cells are attributed with linked information such as bathymetry and bed roughness. Figure 7.1 shows the model domain. A complete description of the DCSM can be found in Gerritsen *et al.* (1995).

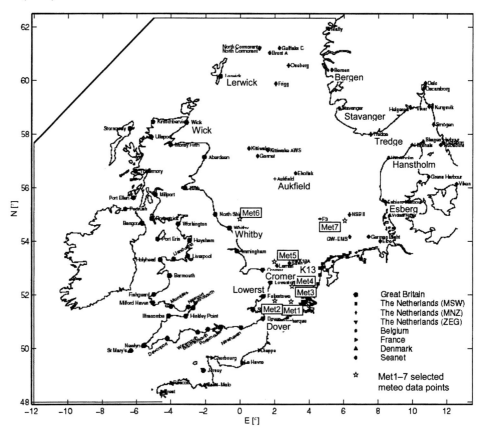

Figure 7.1. Model domain and data points

7.1.2.   HIRLAM

HIRLAM is a numerical weather prediction model maintained by the meteorological authorities of several northwest European countries. In The Netherlands, HIRLAM is maintained by the Royal Netherlands Meteorological Institute (KNMI). The model is used to predict surface pressure, sea surface temperature, wind speed 10 m above the surface and humidity. The lateral boundary values are obtained from a GCM (Global Circulation Model) such as the one operated by ECMWF (European Centre for Medium Range Weather Forecasting). The model is a hydrostatic grid-point model and the resolution in present use is 55 km to 5 km in the horizontal and 16 to 31 levels in the vertical. The coordinate systems used are a rotated latitude-longitude grid horizontally and a hybrid p-sigma system in the vertical. HIRLAM55 (with a 55 km grid) is used in the operational forecasts. The HIRLAM model forecasts provide the meteorological forcings needed by the DCSM to compute the surge.

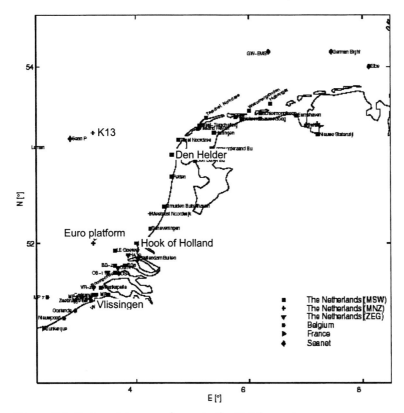

Figure 7.2. Data points on and near to the Dutch coast

7.2.   Data availability and preparation

This study is based on observations and model predictions made in the year 1999. Therefore all the results are applicable to data and model settings for this particular year. However, most of the results also hold for time periods where the involved models are used under settings similar to the year 1999. The locations frequently mentioned in this study are shown in Figure 7.1 and Figure 7.2.

The data are available in different time intervals varying from 10 min to 3 h. In order to use the analysis techniques, it is necessary to select a common base time interval for all the data. A time interval of 1 h is selected on the assumption that it maintains the necessary detail for the data available at 10 min intervals, and does not involve too much interpolation for the data available at 3 h intervals. The data collected for this study include the following:

Observed wind

Observed wind data for K13 platform, Europlatform, Hook of Holland and Vlissingen were obtained from KNMI. The data are available at 1 h intervals as absolute wind magnitude and direction in degrees North.

Predicted pressure and wind

Wind and pressure predicted by the HIRLAM model are available in snapshots at 3 h intervals. The predicted wind is available in the form of East and North components. It is necessary to extract time series data at selected points on the coast and over the sea. Time series data is extracted from the following 12 locations:

K13A (53.22N, 03.22E)	Met2 (52.00N, 02.00E)
Europlatform (52.00N, 03.28E)	Met3 (52.50N, 03.00E)
Hook of Holland (51.98N, 04.12E)	Met4 (53.00N, 03.00E)
Vlissingen (51.44N, 03.60E)	Met5 (53.50N, 02.00E)
Aukfield (56.385N, 02.066E)	Met6 (55.00N, 00.00E)
Met1 (Met1 (52.00N, 03.00E)	Met7 (55.00N, 06.00E)

The points K13A, Europlatform, Hook of Holland and Vlissingen are selected because they have observed wind data and hence it is possible to compute wind prediction errors. Points Met1 to Met7 and Aukfield are selected arbitrarily to study the influence of distance from the coast. Points Met1 to Met5 are located on the narrow part of the sea close to the Dutch and British coasts. Met6 and Met7 are located either side of the widest part of the sea and are mainly selected to study the effect of wind and pressure at farther locations. The data are then interpolated to *hourly* time intervals. The locations are shown in Figure 7.1.

Surge (observed and predicted) and astronomical tide

The *tide (or astronomical tide)* is the astronomical component of the total water level. The *observed tide* is the astronomical component of the water level as computed by harmonic analysis based on long-term observations whereas the *predicted tide* is the water level generated by the DCSM using the astronomical tide components as boundary conditions without using meteorological forcings.

Surge (storm surge) refers to the difference between the actual water level under the influence of meteorological disturbances and the level that would have been attained in the absence of the meteorological disturbance (i.e. the astronomical tide). Storm surge is a consequence mainly of the movement of water under the action of wind stress. A contribution is also made by the hydrostatic rise of water resulting from the lowered barometric pressure. *Observed surge* refers to the difference between the observed water level and observed tide whereas *predicted surge* refers to the difference between the water levels computed with and without meteorological forcings. The *surge prediction error* is computed as the difference between the predicted and observed surge.

The DCSM model does not use observed meteorological data. This is because of the paucity of such data over the North Sea. Instead, the meteorological forcings (wind and pressure fields) are obtained from the HIRLAM model. HIRLAM uses the latest wind and pressure observations in hindcast mode. Hindcasting is done to obtain improved initial conditions for the subsequent operational weather forecast. These hindcast wind and pressure data obtained from the HIRLAM model are referred to as 'predicted'. Predicted surge is therefore the surge computed using the wind and pressure fields obtained from HIRLAM in hindcast mode. In actual operational surge forecasting with DCSM, the wind and pressure fields obtained from HIRLAM in forecast mode are used; such surges are referred as 'forecasted'.

The term *external surge* refers to surges generated by meteorological activities outside the model domain and which enter through the open boundaries of the model. For the domain of the DCSM, the main entry of external surges is the northwest coast of Scotland. However, the DCSM does not assume any external surges in its normal operation.

Both predicted and observed surge and astronomical tide data are available for a number of points around the North Sea at 10 min intervals. Data from 13 stations along the British, Dutch and Scandinavian coasts are selected (see the list in Table 7.1). These data are reduced to hourly intervals. At locations where there is adequate observed surge data, the surge prediction errors are also computed.

Table 7.1. Selected water level data points on the coast

British coast	Dutch coast	Scandinavian coast
Lerwick	Hook of Holland	Esberg
Wick	Vlissingen	Bergen
Whitby	Den Helder	Stavanger
Cromer		Tredge
Dover		Hanstholm

Statistics of the data at Hook of Holland

The basic statistics of the data at Hook of Holland in 1999 are shown in Table 7.2. Figure 7.3 shows the time series of the observed surge, the surge prediction errors, the HIRLAM predicted pressure, and the magnitude and direction of hourly wind data recorded at Hook of Holland in 1999.

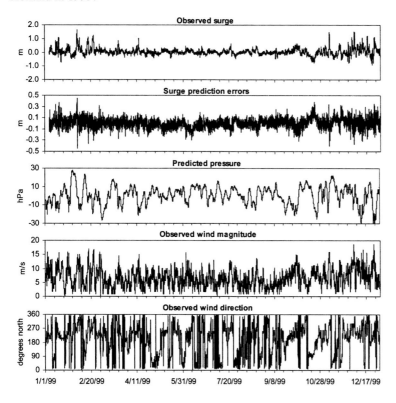

Figure 7.3. Time series of hourly data at Hook of Holland in 1999

Table 7.2. Statistics of selected data at Hook of Holland (all in metres)

Data	Mean	St. Dev.	Minimum	Maximum
Observed surge	0.049	0.250	-0.92	1.66
Surge prediction errors	0.005	0.086	-0.44	0.45
Predicted astronomical tide	0.069	0.632	-0.88	1.54
Observed water level	0.118	0.676	-1.42	2.40

7.3.   Solution approach

The approach followed consists of two parts. The first part is the data relationship analysis, whereas the second part is the actual modelling of prediction accuracy. Thus utilizing the results of the data relationship analysis targeted on confirming or rejecting a number of preliminary hypotheses. Modelling of the surge prediction accuracy is done using appropriate data-driven modelling techniques.

Data relationship analysis

The length and time scales of the wind fields over the North Sea play a crucial role in determining the surge. Comparatively small length scales (~10 km) are important close to the Dutch coast and are less important, perhaps negligible, in the northern part of the North Sea. Similarly, the larger length scales of the wind field (~100 km) are important in the northern part of the North Sea. Connections are sought between the HIRLAM predicted wind at the 55 km grid points over the North Sea and the predicted surge at measurement points along the Dutch coast. It is important to distinguish between the wind normal to the coast and along the coast.

The general progression of tide and surge in the North Sea is typically anticlockwise due to the Coriolis force. Consequently, it is important to look at the connections between the observed surges recorded at sites along the British coast with those along the Dutch coast. It is also necessary to examine connections between the surge prediction errors at these sites. Surge measurements along the British coast are expected to include implicitly significant small length scale wind events near to the measuring locations.

The purpose of the data relationship analysis is to evaluate the content, flow and time dynamics of information regarding the surge and its prediction accuracy with respect to a number of selected parameters. The analysis is done both for the observed surge and the surge prediction errors. The analysis helps to find out how much information contained in the surge prediction errors can be traced back to some of the variables. The test of significance is done using AMI and linear-correlation measures.

The data relationship analysis is done to evaluate the relationship of surges and surge prediction errors along the Dutch coast, in particular at Hook of Holland, with

❑ The predicted wind 10 m above the surface and the air pressure at sea level at grid points of the HIRLAM model.

❑ The total magnitude of the observed wind as well as along and normal to the coast for points adjacent to the coast and for points well away from the coast.

❑ Wind prediction errors at locations where wind measurements are available.

❑ Astronomical tide at locations along the coast around the North Sea.

❑ Observed surge at locations along the coast around the North Sea.

❑ Surge prediction errors at locations along the coast around the North Sea.

A number of hypotheses are put forward regarding the accuracy of the surge prediction at the Dutch coast. It might not be possible to fully confirm or reject some of these hypotheses since the available data is limited. These hypotheses are however used as a starting point for the subsequent analyses.

The hypotheses include:

So far as the accuracy of surge forecasts along the Dutch coast is concerned

H1. The predicted wind and pressure fields away from the Dutch coast but adjacent to another coast are not as significant as predicted wind fields in the middle of the North Sea.

H2. The predicted wind and pressure fields in the northeastern part of the North Sea are less important than those in the southwestern part.

H3. The observed surge along the British coast is significant whereas the observed surge along the German and Scandinavian coasts does not have a significant impact on the surge at the Dutch coast.

H4. Some surge level measurements along the British coast, especially those where there is a complicated local geometry (as for an estuary), are not as significant as sites along the open coast.

H5. The surge prediction errors at key sites along the British coast are significant.

H6. Observed winds near the Dutch coast (within 100 km) are significant (with the wind resolved in directions along and normal to the coast).

H7. The wind and pressure prediction errors at measurement sites near the Dutch coast have a significant impact in surge prediction accuracy (since predicted wind and pressure constitute the forcings for surge prediction).

H8. The previous surge prediction errors along the Dutch coast are important.

H9. The pressure field is not significant in terms of surge prediction errors at the Dutch coast (since its direct contribution to the surge is minimal compared to wind).

H10. There exist external surges across the boundaries of the DCSM - in particular from the Atlantic (data from Bergen, Stavanger, Lerwick, and Whitby will be used to test the hypothesis).

Modelling the prediction accuracy

The second part of the study starts with the selection of the parameters that are best related to the surge prediction errors. This is done based on the results of the data relationship analysis. The main task is to establish a connection between selected predictive parameters and the surge prediction accuracy. Data-driven models are used to model the surge prediction errors in terms of predictive parameters. A proper lead-time is selected to ensure that the surge prediction errors can be forecasted using such a model. Tests are conducted to model and predict the surge prediction errors using artificial neural network models. This serves as a means of evaluating the predictability of the surge prediction errors some time in advance.

If the accuracy of the error forecasts is not satisfactory then the deterministic prediction of errors is not adequate, in which case the problem has to be formulated differently. A possible approach to deal with such a situation is to model the bias and confidence intervals of the surge prediction errors with a neural network. Another alternative is to establish a connection between selected predictors and the surge prediction accuracy using a fuzzy rule-based framework.

7.4.   Surge at Hook of Holland

In this section, the surge data observed at Hook of Holland are analysed in relation to meteorological data at selected locations over the North Sea. The relationship of the surge at Hook of Holland with the surge at other locations around the North Sea is also explored. At first, an analysis of the interaction between astronomical tide and surge at Hook of Holland is analysed using AMI and linear correlation measures.

7.4.1.   Tide and surge

The symmetrical graph in Figure 7.4 shows the autocorrelation and AMI of the surge data at Hook of Holland within lag times of ±48 h. The graph shows that there is a strong serial correlation in the surge data up to a lag time of 6 h. This could be because the wind and pressure also behave in a similar way. Both the AMI and correlation measures show 'bumps' at 12 and 25 h, which are reasonably distinctive. Since 12 and 25 h correspond to the periods of the tidal components, this raises further interest in the relationship between the semidiurnal tidal cycle and the surge. One way to resolve this interest is to analyse the autocorrelation and the AMI of the time series composed of the hourly difference in observed surge. The time series of hourly differences in the observed surge is free from the strong serial correlation in comparison with the actual surge.

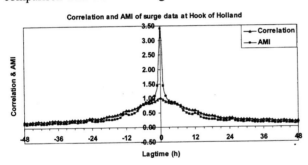

Figure 7.4. Autocorrelation and AMI of the surge at Hook of Holland

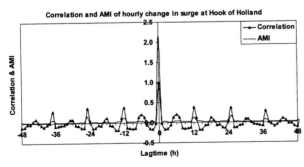

Figure 7.5. Autocorrelation and AMI of the hourly difference in surge at Hook of Holland

Figure 7.5 shows the autocorrelation and the AMI of the hourly differences in surge within lag times of ±48 h. This graph shows that the bumps observed in Figure 7.4 are not only distinct and go as high as a correlation coefficient of 0.5 but also extend to lag times of 37, 50, … hours. Figure 7.6 shows the autocorrelation and the AMI within the astronomical tide data at Hook of Holland. Comparing Figure 7.5 and Figure 7.6 it can be seen that there is a similar pattern in the peaks of the correlation and AMI of the tide and hourly differences in surge observations. This indicates that there is an obvious influence of the tidal cycle on the observed surge. In view of the definitions of the tide and surge such an influence is non-linear.

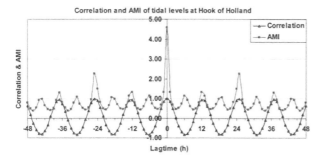

Figure 7.6. Autocorrelation and AMI of astronomical tide at Hook of Holland

7.4.2.   Meteorological variables and surge

Wind and surge

To investigate the relationship between wind (speed and direction) and surge at Hook of Holland, observed wind data at Hook of Holland, K13 and Europlatform are considered. AMI and correlation analyses are applied between the wind speed and surge data up to 48 h.

At first the analysis is made using the absolute magnitude of the wind speed regardless of the direction. Figure 7.7 shows the cross correlation and the AMI between the absolute wind speeds recorded at Europlatform, K13 platform and Hook of Holland and the surge at Hook of Holland. The surge shows a delayed response, peaking at a lag time within 6 to 9 h for the wind at Hook of Holland and 12 to 15 h for that of K13 platform, which is the farthest among the three locations. The figure also shows that the strength of the relationship decreases weakly with distance from the coast.

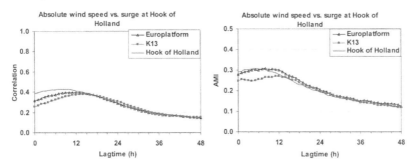

Figure 7.7. Relationship between absolute wind speed and surge at Hook of Holland

In order to investigate the importance of wind direction, the analysis is carried out using wind components resolved along and towards the coast. The average orientation of the Dutch coastline near Hook of Holland is approximately N40^{0}E. The sign convention used here is such that winds from the southwest assume positive sign along the coast and winds from the northwest assume a positive sign towards the coast (see Figure 7.8). Subsequently the wind data at Hook of Holland are resolved into these two directions and the nature of their relationship with the observed surge at Hook of Holland is analysed.

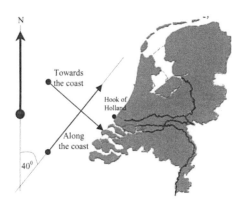

Figure 7.8. Orientation of the coastline and sign convention of wind components along and towards the coast

Wind speed towards the coast: Figure 7.9 shows the correlation and the AMI between the wind component towards the coast and the surge at Hook of Holland. The results show that wind component towards the coast has an immediate effect on the surge and has a positive correlation. The faster the wind the higher the surge, and if the direction of the wind is reversed the surge is negative, which agrees with the expectation. There is no visible difference in the response time between all three stations. The AMI and correlation of the wind component towards the coast are stronger than that of the absolute wind speed. Both the AMI and correlation values drop steadily in 15 h.

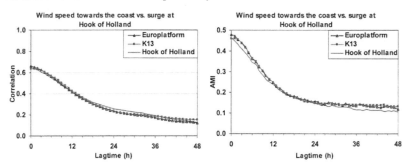

Figure 7.9. Relationship between wind speed towards the coast and surge at Hook of Holland

Wind speed along the coast: Figure 7.10 shows that the response of the surge to the wind component along the coast is delayed, unlike the response from the component towards the coast. The peaks of the AMI and correlation measures lie between lag times of 15 and 18 h in all three cases. The response is consistently weaker than that for the wind towards the coast. Also, the correlation coefficients always suggest that southwesterly winds result in a delayed positive surge whereas northeasterly winds are associated with a delayed negative surge. This is consistent with the physical explanation that the water is pushed towards the coast by the Coriolis effect.

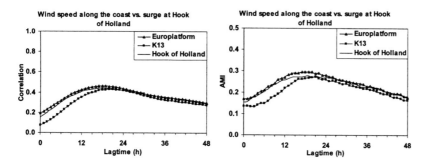

Figure 7.10. Relationship between wind speed along the coast and surge at Hook of Holland

One important observation here is that the distinctive response of the resolved wind components and the higher AMI values with the surge compared to that of absolute wind speed suggest that future analyses of this sort have to be conducted with the resolved wind components rather than the absolute magnitude.

Pressure and surge

The relationship between the atmospheric pressure at sea level and the surge is explored using predicted pressure data since measured pressure data are not available with the required temporal and spatial resolution. Figure 7.11 shows the autocorrelation and the AMI of the predicted pressure at Hook of Holland. The fact that the autocorrelation stays high for a long period (well above 0.75 up to 24 h) suggests a persistent behaviour.

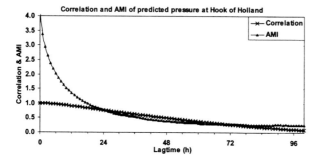

Figure 7.11. Autocorrelation and AMI within the predicted pressure at Hook of Holland

The predicted pressure at Hook of Holland, Europlatform, Aukfield and Met6 is used to study the relationship between the pressure and the surge. Both the cross correlation and AMI plotted in Figure 7.12 show that the surge reveals a delayed response to the pressure no matter how close to the coast the pressure prediction point is. The response time varies from 9 h for the pressure at Hook of Holland to 18 h for the farthest location, Met6. This gives an insight into the length and time scales involved between pressure and surge. An important point here is that the surge at Hook of Holland is related more strongly to the pressure at Aukfield than the pressure both at the closer locations Hook of Holland and Europlatform and the farther location Met6. Also the farthest location Met6 relates better than the nearest locations. A visible pattern here is that the response associated with pressure is stronger for locations far from any coast and possibly in the deep sea. The fact that the correlation is always negative means that low pressure causes a positive surge and vice versa, which agrees with the known physics of the situation.

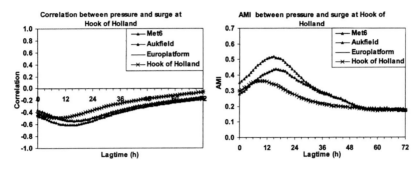

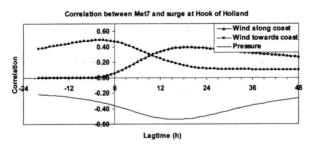

Figure 7.12. Relationship between pressure at four locations and surge at Hook of Holland

Comparing the maximum AMI values with those of the wind speed, it can be seen that pressure is much more important than initially expected. From a theoretical point of view, naturally observable magnitudes of pressure are not capable of causing so much response in comparison to other forces acting on the water body such as the wind shear. The possible explanation for the presence of a strong correlation between pressure and surge might be explained by the inherent relationship between pressure difference and wind (speed and direction).

Wind and pressure at the northeastern part of the North Sea

To study the significance of the meteorological variables on the northeastern part of the North Sea, the predicted wind and pressure at Met7 close to NSB II (Figure 7.1) are used. The correlation and AMI (Figure 7.13 and Figure 7.14) show strong responses especially for wind speed towards the coast and the pressure. Negative lag times were also used in this analysis. It is interesting to notice that the wind component towards the coast, which caused an immediate response for the case of all the other locations, has a peak on the left-hand side of the vertical axis in the case of Met7. It appears that the wind comes after the surge, which contradicts the physical relationship between the wind and the surge. The implication is that the meteorological variables on the northeastern part of the North Sea are less important. It is therefore more likely that the correlation is a result of the general eastward movement of the weather pattern.

Figure 7.13. Cross correlation between meteorological variables at Met7 and surge at Hook of Holland

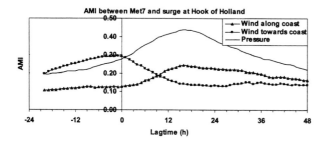

Figure 7.14. AMI between meteorological variables at Met7 and surge at Hook of Holland

7.4.3.   Relationship between surge at Hook of Holland and other locations

Here, the strength of the relationship and the time dynamics involved between the surge at Hook of Holland and other locations around the North Sea is considered. The selected data points are grouped into three: the British coast, on/near the Dutch coast and the Scandinavian coast. The correlation and AMI within ±48 h are analysed. Reverse lag times are intentionally included since the direction of information propagation is also important.

Locations on the British coast

Locations on the British coast are particularly important since the inherent direction of the tidal movement is counterclockwise. Figure 7.15 and Figure 7.16 respectively show the cross correlation and the AMI between the surge along the British coast (Lerwick, Whitby, Cromer and Dover) and at Hook of Holland. The result from both measures is consistent in that the lag times corresponding to a maximum relationship agree with the direction of the tide. It is also important to notice that Whitby is significant since it shows a higher relationship, yet provides a lead-time of up to 7 h if forecasting is intended. Surges at Lerwick and Hook of Holland show a separation of +20 h. This relatively low yet distinctly peaked relationship shows that the effect of external surges from the northern boundary cannot be ruled out. However, only one year of data would have too few such events to confirm this. Dover shows a double peaked relationship at −1 h and +4 h. This indicates a bi-directional flow of information, which might be attributed to its geographic proximity to Hook of Holland and the narrow width of the English Channel.

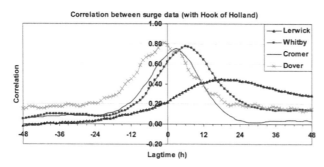

Figure 7.15. Cross correlation between surge at the British coast and Hook of Holland

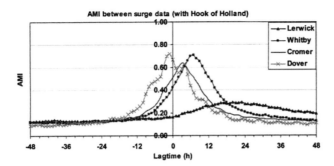

Figure 7.16. AMI between surge at the British coast and Hook of Holland

Locations on /near the Dutch coast

The locations Vlissingen, K13 platform, Den Helder and Esbjerg Harbour (not on the Dutch coast) are in this group. As may be expected from the proximity of these locations to Hook of Holland, Figure 7.17 and Figure 7.18 show that the information from these locations has peaks near the zero hour lag time. Also all locations except Esberg show very strong responses compared to locations on the British coast.

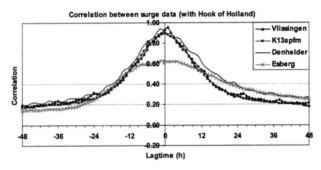

Figure 7.17. Cross correlation between surge at the locations on/near the Dutch coast and Hook of Holland

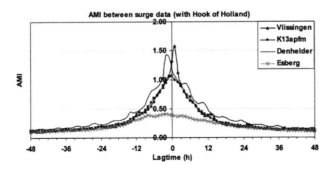

Figure 7.18. AMI between surge at the locations on/near the Dutch coast and Hook of Holland

Locations on the Scandinavian coast

The points selected on the Scandinavian coast are Hanstholm, Tredge, Stavanger and Bergen. Figure 7.19 and Figure 7.20 show that generally these locations show a weaker relationship with Hook of Holland compared with the other locations. The relationship with Hanstholm and Tredge has peaks at negative lag times showing that, on average, these locations receive information from Hook of Holland and not vice versa.

The correlations with Bergen and Tredge show two peaks, one with a negative lag and one with a positive lag. This indicates that there are two mechanisms, moving in two different directions. The physical mechanism behind this is not well understood. It could be that external surges show up on both sides of the northern North Sea in addition to the general counterclockwise propagation of the tides.

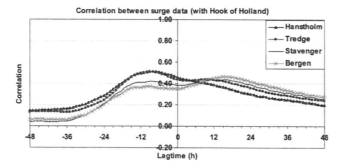

Figure 7.19. Cross correlation between surge at the Scandinavian coast and Hook of Holland

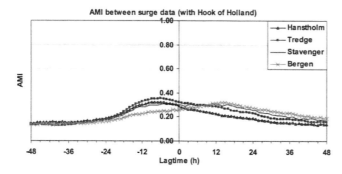

Figure 7.20. AMI between surge at the Scandinavian coast and Hook of Holland

Summary

The maximum values of the AMI and the correlation coefficient between the surge at Hook of Holland and other locations are summarized in Table 7.3. The values of the maximum AMI and correlation measures show the relative strength of the relationship between the surge at Hook of Holland and other locations. The lag times corresponding to the maximum AMI show the time dynamics involved with the movement of the surge.

Table 7.3. Relationship between observed surge at Hook of Holland and other locations

Station	Maximum AMI	Maximum correlation	Lagtime (h) for maximum AMI (*in phase with correlation*)
Lerwick	0.29	0.44	+20
Whitby	0.71	0.77	+7
Cromer	0.64	0.75	+3
Dover	0.72	0.81	-1, +4
Vlissingen	1.58	0.96	+1
K13A	1.06	0.90	-1
Hook of Holland	3.41	1.00	0
Den Helder	1.42	0.94	-2
Esberg	0.41	0.62	-2
Hanstholm	0.32	0.51	-5, -9
Tredge	0.35	0.51	-6
Stavanger	0.31	0.44	+12, -4
Bergen	0.32	0.46	+15

The effect of the tide on the surge has already been established. The time dynamics of the surge however do not exactly coincide with the tidal cycle in the North Sea. This and the presence of double peaks in some of the relationships, such as Dover, Hanstholm and Stavanger, suggest that there are another important phenomena, such as the dominant direction of the wind, affecting the information flow associated with the surge. One of the peaks with Stavanger is at +12 h, which could be an indication for a possible effect of external surges from the North Atlantic. The pattern of the relationship with Bergen in (Figure 7.19 and Figure 7.20) is very similar with that of Lerwick (Figure 7.15 and Figure 7.16). This is particularly true for the right-hand side of the curves. This can be a further evidence for the presence of external surges. If the tide were the only process driving the surge, Bergen would have only one peak on the left hand side (with a negative lag time). The correlation between the surges is always positive for a wide range of lag times, even for locations far from Hook of Holland.

7.5. Surge prediction errors at Hook of Holland

The surge prediction errors at the Dutch coast, particularly at Hook of Holland, lie at the centre of all the analyses in the following sections. The analysis made on the surge at Hook of Holland has already given some insight into the relationship and interaction between various variables and the surge. This interaction is solely a function of the physical system. Here, a more or less similar analysis is made with the surge prediction errors. The interaction between other variables and the surge prediction errors however is not driven by the physics alone. There is also the influence of the model. Since the prediction errors are assumed to represent the gap between the model and the physical system, it is necessary to interpret the results accordingly. The analysis is intended to relate the surge prediction errors at Hook of Holland to the observed and predicted meteorological variables and the surge prediction errors at other locations.

Before seeking relationships between the surge prediction errors at Hook of Holland and other variables, it is necessary to examine the information contained within the time series of the error at Hook of Holland itself. Figure 7.21 shows the autocorrelation and the AMI of the surge prediction errors at Hook of Holland within ±48 h. The symmetrical plot bears a striking resemblance with Figure 7.5. The same conclusions made there also apply here except that here it means that the tidal cycle contributes to errors in surge prediction. This is particularly important since a correlation coefficient of 0.5 corresponding to 25 h lag time means that the effect of the tide on the error is significant. It is also an indication that so much information is available if it is intended to forecast the errors, say, 25 h in advance.

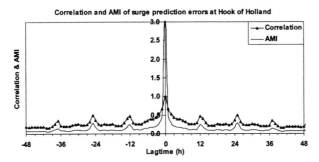

Figure 7.21. Autocorrelation and AMI of surge prediction errors at Hook of Holland

7.5.1.   Meteorological variables and surge prediction errors

Wind components and surge prediction errors

Observed wind data at Hook of Holland and K13 platform are used to investigate the relationship between the wind speed and surge prediction errors at Hook of Holland. The analysis is made using the components along and towards the coastline since it is already established that it is better than using the absolute wind magnitude.

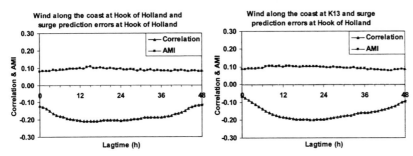

Figure 7.22. Relationship between wind along the coast and surge prediction errors at Hook of Holland

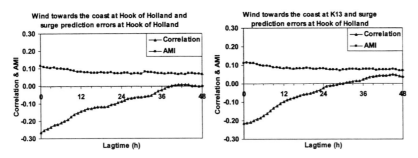

Figure 7.23. Relationship between wind towards the coast and surge prediction errors at Hook of Holland

Wind speed along the coast: The relationship with the wind along the coast and surge prediction errors at Hook of Holland is shown in Figure 7.22 for wind data at Hook of Holland and K13 platform. Even though it is weak, the response is distinct in that it has a delayed peak between 12 and 24 h in both cases. The correlation coefficient is consistently negative showing that faster positive winds along the coast are associated with negative errors, causing underestimation of the surge, and vice versa.

Wind speed towards the coast: The relationship between the surge prediction errors and wind towards the coast shown in Figure 7.23 has some differences from that along the coast. There is an immediate response of the surge errors to the wind towards the coast. The relationship decreases steadily with lag time. The correlation coefficient is negative until it changes sign well after 24 h in both cases.

Wind prediction errors versus surge prediction errors

Since wind is the main driving factor for the surge, it is important to see how the wind prediction errors are related to the surge prediction errors. The resolved components of the wind prediction errors at Hook of Holland and K13 platform are used since both observed and predicted wind data are available at these locations. The correlation and AMI plots shown in Figure 7.24 and Figure 7.25 show that the wind prediction errors both along and towards the coast show some patterns but the relationship is rather weak.

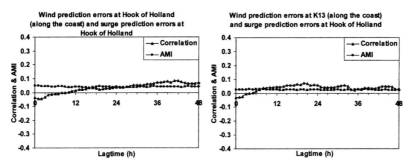

Figure 7.24. Relationship between wind prediction errors along the coast and surge prediction errors

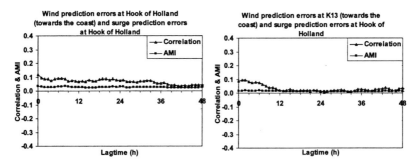

Figure 7.25. Relationship between wind prediction errors towards the coast and surge prediction errors

Predicted pressure and surge prediction errors

The significance of the relationship between pressure and surge prediction errors is investigated using the predicted pressures at Hook of Holland, K13 platform, and Met6. The results are plotted in Figure 7.26. The graphs show a weak but clear response that is delayed between 12 and 24 h. The correlation is consistently positive unlike the correlation between pressure and surge, meaning that high pressure is associated with positive errors (overestimated surge prediction) and vice versa. Also the predicted pressure at the farther location Met6 relates better than the pressure at locations closer to Hook of Holland.

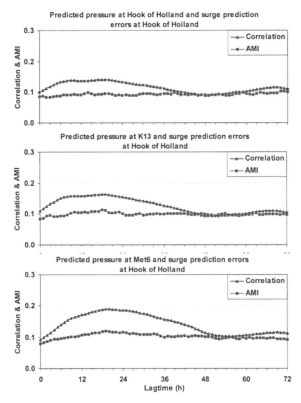

Figure 7.26. Relationship between predicted pressure and surge prediction errors at Hook of Holland

Wind and pressure at the northeastern part of the North Sea

To study the contribution of the meteorological variables on the northeastern part of the North Sea, the predicted wind and pressure at Met7 are used once again. Cross correlation and AMI analyses are carried out between the meteorological variables at this location and the surge prediction errors at Hook of Holland. The results showed weak correlation and AMI (see Figure 7.28 and Figure 7.27). This and the contradictory direction of information flow obtained in §7.4.2 to that of the cause and effect relationship between wind and surge, suggest that the wind and pressure in the northeastern part of the North Sea are not of much importance in forecasting the surge prediction accuracy at Hook of Holland.

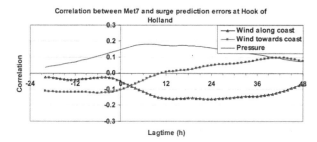

Figure 7.27. Cross correlation between Met7 and surge prediction errors at Hook of Holland

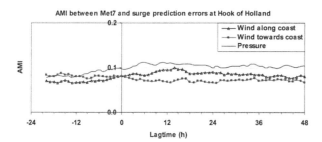

Figure 7.28. AMI between Met7 and surge prediction errors at Hook of Holland

7.5.2.   Surge prediction errors at Hook of Holland and other locations

As in the case of the observed surge, the relationships between errors in the predicted surge at other locations with that at Hook of Holland are analysed in three groups: the British coast, on/near the Dutch coast and the Scandinavian coast.

Locations on the British coast

Figure 7.29 and Figure 7.30 show that the surge prediction errors on the British coast are related to those at Hook of Holland with varying degrees and lag times, which are generally a function of the distance involved. Whitby is still important, trading off between relatively higher information content and a longer lag time (+7 h). But all the locations on the British coast could be important in forecasting the errors at Hook of Holland.

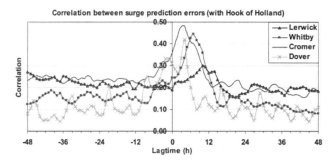

Figure 7.29. Cross correlation between errors surge prediction at the British coast and Hook of Holland

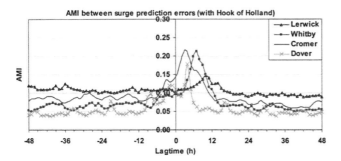

Figure 7.30. AMI between errors surge prediction at the British coast and Hook of Holland

Locations at/near the Dutch coast

As can be seen in Figure 7.31 and Figure 7.32, the locations at/near the Dutch coast generally have highly related surge error patterns with Hook of Holland. However, the time of separation is nearly insignificant which compromises their usability for error forecasting purposes. Vlissingen and Den Helder are the locations most related to Hook of Holland, obviously because of their geographical proximity.

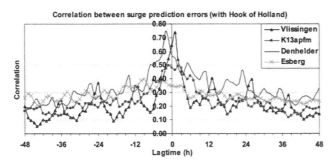

Figure 7.31. Cross correlation between surge prediction errors at/near the Dutch coast and Hook of Holland

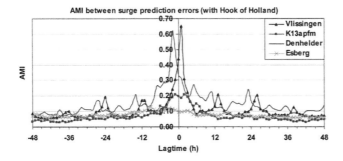

Figure 7.32. AMI between surge prediction errors near the Dutch coast and Hook of Holland

Locations on the Scandinavian coast

The relationship with the surge prediction errors with the Scandinavian coast is the weakest, as shown in Figure 7.33 and Figure 7.34. Also the lag times corresponding to maximum information are negative, showing that these locations are receptors of information.

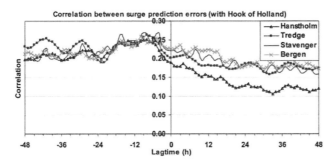

Figure 7.33. Cross correlation between surge prediction errors at Scandinavian coast and Hook of Holland

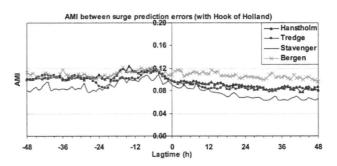

Figure 7.34. AMI between surge prediction errors at Scandinavian coast and Hook of Holland

Summary

Table 7.4. Relationship between surge prediction errors at Hook of Holland and other locations

Station	Maximum AMI	Lagtime for maximum AMI (h)
Lerwick	0.148	+10
Whitby	0.212	+7
Cromer	0.215	+2,3
Dover	0.178	+4
Vlissingen	0.653	+1
K13A	0.210	-1
Hook of Holland	3.378	0
Den Helder	0.618	-2
Esberg	0.128	-9
Hanstholm	0.123	-14(-8)
Tredge	0.118	-7(-8)
Stavanger	0.108	-8
Bergen	0.122	-16(-10)

The peak AMI values and the corresponding lag times between surge prediction errors at all the locations and Hook of Holland are summarized in Table 7.4. The magnitude of the lag times and their algebraic sign shows that the dominant direction of information flow concerning the surge prediction errors is counterclockwise. Most of the Scandinavian locations showed competitive secondary peaks.

7.6. Summary of the data relationship analysis

The preliminary data analysis conducted with the observed data and model simulation for the year 1999 has revealed important points regarding the time dynamics and length scales involved in the interaction between the surge data observed at various locations around the North Sea, the interaction between meteorological variables and the surge and surge prediction errors. Some of these results are open to speculation since, among other things, the analysis is done with data containing limited extreme events, and also since there might be alternative explanations to some of the conclusions drawn from the analysis with the surge data and the prediction errors.

The surge data at Hook of Holland has within it a periodic component. Surge by definition is a weather driven phenomenon and not periodic since the driving forces are not periodic. This however can be explained by the fact that the driving forces act on the sea at different tidal states and the surge response also bears the effect of those tidal states. In addition, the surge prediction errors at Hook of Holland show a periodic behaviour that is consistent with the tidal cycle. The most dominant cycle has a period of 25 h and the next has a period of 12 h (it has to be noted that AMI and correlation analyses can determine the period only to the nearest time step). This however, is an indication that the model did not properly incorporate the interaction between the surge and tide. From another perspective, the relationship between surge prediction errors and the tidal cycle indicates a systematic predictability of part of the surge prediction errors with a separate model.

The influence of the tidal cycle further extends to the relationship between the surges observed at locations around the North Sea. It is also revealed in the relationship between the surge prediction errors at these locations. This is because the direction of information flow follows the characteristic direction of the tidal movement in the North Sea. However, the travel times of the information do not exactly coincide with the difference in tidal hours between these locations, which indicates that there are other processes involved in the information travel. One possible explanation can be the presence of a dominant wind direction. But to confirm this it is necessary to analyse the pattern of meteorological events over the Sea over a rather longer time domain.

The response of the surge at Hook of Holland to the meteorological data sampled at different locations shows the time and length scales involved in the interaction. One of the most important observations made is the importance of using the wind speed resolved into along and towards the coast components rather than in absolute terms. The two components relate to the surge in entirely different ways. The surge responds immediately to the component towards the coast whereas the response to the component along the coast comes more than half a day later. The other observation is the stronger than expected response of the surge to the predicted pressure. The negative sign of the correlation coefficient agrees with the theoretical explanation. The high value of the correlation, however, might not imply a direct cause and effect relation. Instead it might be an indirect or secondary effect such as the pressure affecting the wind and the wind affecting the surge. The magnitude and response time of the surge to the pressure varies visibly with the distance of the pressure data point from Hook of Holland. The response is apparently stronger for pressure data taken farther from any coast.

The wind and pressure data also show patterns in their relationship with the error in the predicted surge at Hook of Holland. In particular, the consistent sign of the correlation coefficient in each of the relationships is an indication that there is a typical combination of wind directions and pressure that contributes to extreme errors. For instance, northwesterly winds are associated with an underestimation of the immediate surge predictions. The faster they are the greater the underestimation becomes. In an ideal model, there should be no relationship between any of the forcings used by the model and the residual errors. If there is, it means that the available information is not fully used by the model. Knowledge of the presence of such a relationship can, however, be used in improving the prediction/forecast model, in data assimilation, or error forecasting.

The relationships between wind prediction errors and surge prediction errors were too weak to find any identifiable pattern compared to the relationship with the magnitude of the observed wind and thus were inconclusive. These results were against the initial expectations that gave more importance to the effect of wind prediction errors on the accuracy of surge prediction. This does not mean that the error in wind prediction is not important; its importance in determining the accuracy however, is not as dominant as other factors.

There is also some evidence for the influence of external surges from the North Atlantic. The reason for this is because information flows to Hook of Holland from locations on the British coast as far as Lerwick. The relationships with Stavanger and Bergen, which are close to Lerwick but on the other side of the sea, are double peaked where the second peak has a positive lag time suggesting that there is a flow of information from these locations to Hook of Holland.

Using the data relationship analysis, the preliminary hypotheses stated in §7.3 are re-evaluated as follows:

H1.   Wind and pressure are certainly related to the surge and its prediction errors. Predicted pressure on the deep sea is more important than near the coast.

H2.   True.

H3.   Generally true. However, Bergen and Stavanger seem important in a different way since these data suggest the presence of external surges from the North Atlantic boundary of the model.

H4.   This hypothesis has been used in the selection of the data points on the British coast and therefore has not been verified.

H5.   True. Both the magnitude of the relationship (correlation and AMI) and the predominant direction of information flow are favourable.

H6.   The effect of distance is not important. However, resolving the wind to components along and towards the coast was found to be important and gave more important insight than using the absolute magnitude.

H7.   The wind prediction errors do not show a strong relationship with the surge prediction errors, so nothing in particular can be said. The effect of errors in pressure prediction could not be confirmed since observed pressure data was not available.

H8.   True. Previous surge prediction errors are certainly important. Using the previous errors it was possible to establish the presence of periodic components in the predicted surge as well as in the surge prediction errors. This was an indication of the influence of the tidal cycle in the sea.

H9.   False. Pressure is related to the surge prediction errors even more than initially anticipated.

H10.   There is evidence suggesting the presence of external surges from the North Atlantic since there is a flow of information from the surge data at Lerwick, Bergen and Stavanger to those at Hook of Holland.

7.7.   Forecasting surge prediction errors using neural networks

Any data-driven modelling technique can in principle be used for predicting errors of hydrodynamic models. In this respect, ANN models have been applied in the recent literature (see, for example, Babovic *et al.*, 2001; Abebe & Price, 2000). Other data-driven modelling techniques such as local linear models based on chaos theory have also been applied in coastal problems (see, for example, Frison *et al.*, 1999; Frison, 2000; Velickov, 2002; Solomatine *et al.* 2001). However, such techniques demand much more data than is used in this study. Due to these reasons, neural network models are used here to model the relationship between selected parameters and the surge prediction errors at Hook of Holland.

The data relationship analyses showed that surge prediction errors at Hook of Holland are related with varying degrees and patterns to the considered parameters. The relationship with parameters such as wind speed towards the coast shows an immediate response whereas the relationship with wind speed along the coast and surface pressure shows a delayed response. The relationship with surge errors at other observation points shows a periodic pattern with some phase lag depending on the location of the point in relation to Hook of Holland. The degree of the relationship between a parameter and the surge prediction errors is an indication of the predictability of the surge prediction errors with a data-driven model using that particular parameter.

While it is important to know the presence of a predictable pattern in the surge prediction errors, the question that still remains is to what degree is it predictable? This question cannot be answered based only on the results of the data relationship analysis. One reason for this is that there is an inherent dependency among some of the predictive parameters, which indicates of the presence of some redundancy. In other words, if all the related parameters are used in a data-driven model to predict the errors, their predictive capability will not be additive. If a hypothetical function $Prd(X)$ is defined to represent the predictability of the surge prediction errors using a parameter X, then the following relationship likely holds:

$$Prd(X_1, X_2, ..., X_N) \leq Prd(X_1) + Prd(X_2) + ... + Prd(X_N) \tag{7.1}$$

Intuitively, the left and right hand sides of this relationship become equal if and only if all the predictive parameters are independent with respect to their contribution to the prediction accuracy. The best way to determine to what extent the errors can be predicted is to test what portion of the relationship a data-driven model can catch. This is not a trivial problem since, even using the same set of predictive parameters as input, different data-driven modelling systems can result in models with varying accuracy. Even the same data-driven modelling system, such as an ANN modelling system, can end up with models of varying accuracy since their structure can vary and the resulting models can be sub-optimal.

7.7.1.   Forecast horizon

In most forecasting problems, forecast accuracy is higher for shorter horizons. The findings of this study will be extended and applied in forecast mode and hence a feasible forecast horizon has to be set in the process of selecting predictive parameters. A horizon of 6 h is set to forecast the surge prediction errors. If the relationship of a parameter to the surge prediction errors shows an immediate response, then the contribution of this parameter to the accuracy of the ANN model depends on the forecast horizon. That is not the case for parameters showing a response delayed by more than 6 h.

7.7.2.   Selection of predictive parameters

The parameters considered to develop an ANN model of the surge prediction errors are composed of meteorological data (wind and pressure) and past surge prediction errors at Hook of Holland and other locations.

Sea surface pressure: The predicted surface pressure at all the locations tested in the data relationship analysis shows a delayed response on the surge prediction errors. The delay in all the cases is considerably more than 6 h and is suitable for a forecast horizon of 6 h. Therefore, the data point with the maximum AMI, which in this case is Met6, is selected. The lag time corresponding to the maximum AMI value is t-18 hours.

Wind speed: The observed wind data at Hook of Holland is considered since it is available. The patterns of the relationship of wind speed along and towards the coast to the surge prediction errors are different. The wind component along the coast has a delayed response with a lag time corresponding to the peak AMI at t-15 hours. The wind component towards the coast has an immediate response so it is more effective to use a lag time as low as possible. Since the prediction horizon is set to 6 h, the highest AMI that can be used without compromising the prediction accuracy is at t-6 hours.

Past surge errors at Hook of Holland: The surge prediction errors at Hook of Holland show some serial correlation. But at a lead-time of 6 h this correlation is weak compared to the periodicity in the time series that appears at 12, 25, 37... hours. At this stage, the surge prediction errors at t-12 and t-25 hours are considered.

Surge errors at other locations: Surge prediction errors at Dover, Den Helder, Vlissingen and Whitby are highly related to those at Hook of Holland with respectively decreasing orders of the maximum AMI. At Dover and Vlissingen the separation time is 1 h and 2 h with Hook of Holland, which is not suitable for a horizon of 6 h. Den Helder is a receiver of information from Hook of Holland. Whitby shows a peak AMI with Hook of Holland for a separation time of 7 h, so it is considered at t-7 hours despite the fact that it is not the best related parameter compared with the other three locations. In fact, if a longer prediction horizon is intended, even the farther locations on the Scottish coast could be considered. A logical reason for testing observations on the British coast is the possibility of incorporating additional information from the other side of the North Sea, despite the fact that these data could be influenced by local meteorological disturbances and local physical conditions.

7.7.3.   Tests on neural network error forecasts

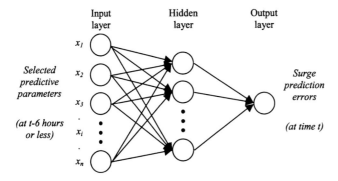

Figure 7.35. Neural network structure for forecasting surge prediction errors

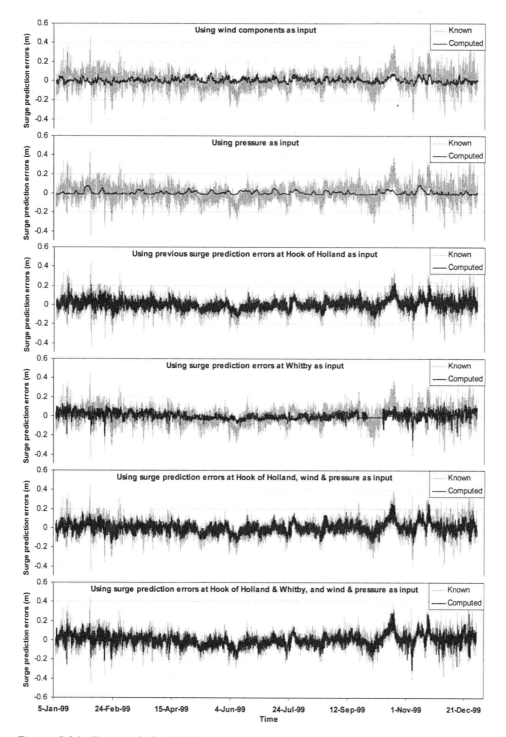

Figure 7.36. Errors of the DCSM model (known) and errors forecasted by ANN using different sets of input data (computed)

The ANN model selected here is the MLP type network with a back-propagation (BP) training algorithm. The structure of the network is shown in Figure 7.35. The tests are done in two stages: first each selected parameter is used as input separately. Then all parameters are used together. This helps to evaluate the relative importance of each selected parameter and their collective importance in forecasting the surge prediction errors.

At first the whole data for the year 1999 at a time interval of 1 h is used for training. The data is later split into two: the first 6 months (January to June) is used for training and the remaining 6 months (July to December) is used to verify the performance of the trained neural network model.

The whole data used for training

ANN models with different combination of input data were trained using the whole data for training. The DCSM errors and the ANN forecasted errors are plotted in Figure 7.36. The training errors are summarized in Table 7.5.

Table 7.5. Training errors of ANN error models using different sets of input data

No	Input to ANN	RMSE (m)
0	DCSM (without ANN error model)	0.087
1	Wind	0.083
2	Pressure	0.084
3	Surge prediction errors at Hook of Holland	0.070
4	Surge prediction errors at Whitby	0.077
5	Combined (1, 2, 3)	0.066
6	Combined (1, 2, 3, 4)	0.059

Data split for training and verification

For the split data analysis, an ANN with five input nodes (pressure, the two wind components, and surge prediction errors at Hook of Holland and Whitby) is used since it shows the best performance in the above tests. The time series of known and predicted errors are shown in Figure 7.37. Although the results are plotted on the same set of axes, the first half is the training performance whereas the second half is the verification performance.

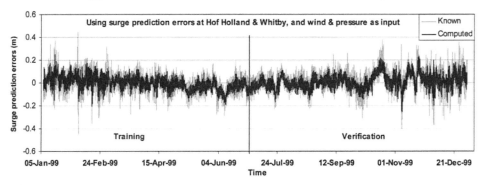

Figure 7.37. Errors of the DCSM model (known) and errors forecasted by ANN (computed)

7.7.4.   Accuracy of error forecasts

A comparison is made between the accuracy of the surge predictions of the DCSM with and without the ANN error forecast model. Different error measures and the 90% confidence intervals are used. The confidence intervals are computed using the procedure described in §7.8 These measures are also computed on monthly basis in order to assess seasonal variations in the surge prediction accuracy.

Table 7.6. Accuracy (RMSE in m) of DCSM, ANN with data split for training and verification

Data set	DCSM	+ANN
All year data	0.086	0.063
January to June (training)	0.083	0.059
July to December (verification)	0.090	0.067

Table 7.7. Confidence intervals of DCSM prediction errors and residuals of ANN error forecasts

Data set	MAE (m)		90% interval (±) (m)	
	DCSM	+ANN	DCSM	+ANN
All year data	0.0661	0.0466	0.1396	0.1004
January to June	0.0637	0.0439	0.1332	0.0932
July to December	0.0684	0.0492	0.1452	0.1055

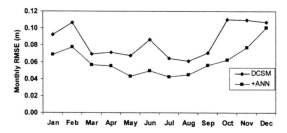

Figure 7.38. Monthly RMSE of the surge prediction of DCSM and the ANN error forecasts

The results showed that the predictive parameters used to forecast surge prediction errors at Hook of Holland catch different patterns of the error. The meteorological parameters (wind components and pressure) tend to capture the pattern in the errors that appears to be meteorologically induced. The past errors at Hook of Holland capture the periodicity thus providing information different from wind and pressure. The errors at Whitby also show a performance comparable to past errors at Hook of Holland. However, the data at Whitby consists of continuous patches of missing data in September and October. Figure 7.38 shows that the use of an ANN model improves the surge prediction accuracy in all the months. The RMSE computed both on the whole data and on monthly basis is reduced consistently by predicting the errors.

The following observations can be made based on the overall evaluation of the results obtained from ANN models used for direct forecasting of surge prediction errors:

❑ All of the selected parameters contribute to the error forecast model and are related to the surge prediction errors in one way or another confirming the results of the AMI analysis.

❑ Direct forecasting of errors with ANN models can only handle part of the surge prediction accuracy. Considerable residual errors still remain leaving a sizable uncertainty in the predicted surge. Therefore, alternative solutions have to be devised to deal with the surge prediction accuracy or at least the remaining residual from the ANN model. One possible alternative is test other predictive parameters such as wind and pressure sampled at points other than the ones considered or derived forms of predictive parameters such as pressure difference between two locations, or differences in time. Another alternative is to try forecasting the surge prediction accuracy in some other form that direct error forecasting. The latter alternative is explored in the following sections.

❑ There is a seasonal variation in the accuracy of the surge prediction, which is better in summer months than in the winter months.

7.8.   Forecasting surge prediction accuracy using neural networks

From the results obtained so far, it has been established that forecasting surge prediction errors with an ANN model does leave a rather significant portion of the errors unaccounted for. This is an indication that, at least with the selected set of predictive parameters, this is as far as an ANN error forecast model can go. This case study presents an example in which error forecasting with a separate model is not sufficient to manage the prediction uncertainty. It is necessary to devise a way to account for both the recoverable and the non-recoverable portion of the surge prediction errors from the DCSM.

A possible solution for this is to develop a neural network model that can forecast the prediction accuracy in a form different from direct error forecasting. One possible alternative is to forecast the confidence intervals of prediction errors. In addition to error measures such as root mean square error, the accuracy of model predictions can be evaluated using confidence intervals. A confidence interval is the range within which model predictions fall with a predefined probability, which gives information regarding the accuracy of predictions that is different from that provided by error measures.

Computation of bias and confidence bounds

At first the bias and confidence intervals have to be computed to see if there is a pattern that can be related to the predictive parameters. Confidence intervals are computed based on the percentiles below or above selected probabilities from the distribution. In this study, the confidence limit is set to 90%. Confidence intervals can be computed using signed or absolute errors. In case of using absolute errors, the 90% confidence interval is the magnitude of the error that exceeds 90% of the errors in the sample population. When signed errors are used, two values are used to mark the 90% confidence interval. The lower bound is the value that exceeds 5% of the errors in the population whereas the upper bound exceeds 95% of the errors. That means 90% of the errors fall within these two bounds. The computation of confidence intervals in both cases is illustrated in Figure 7.39(a) and (b). Here the confidence bounds are computed using signed errors.

The bias of the prediction errors is a measure of the tendency of over- or under-prediction by a model and is defined as:

$$Bias = \frac{1}{N}\sum_{i=1,N}(P_i - O_i) = \bar{P} - \bar{O} = \bar{E} \tag{7.2}$$

where P and O are the predicted and observed values respectively, E is the error and N is the number of predicted values.

The bias will be positive if the model is generally over-predicting and negative if the model is under-predicting the values in comparison to measurements.

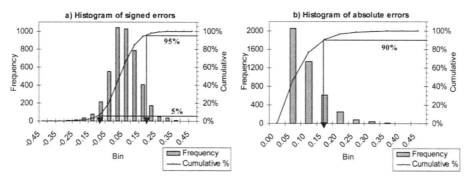

Figure 7.39. Computation of 90% confidence intervals for signed and absolute errors

7.8.1.   Confidence bounds in terms of predictive parameters

To compute the pattern of the bias and confidence intervals with respect to a selected predictive parameter, the available range of the predictive parameter is divided into bins. Traditionally, bins are constructed in such a way that they have equal intervals. That is not feasible here since the samples corresponding to the extreme values of the predictive parameters are very sparse. The bins are therefore divided in such a way that roughly equal number of samples fall in each bin meaning that the bins will have variable sizes.

For each bin the following four values are computed:

❑   Mean value of the predictive parameter,

❑   Bias of the surge prediction error,

❑   Lower bound of the 90% confidence interval, and

❑   Upper bound of the 90% confidence interval.

The bias and lower and upper bounds of the 90% confidence interval of the surge prediction errors are used as indicators of the prediction accuracy. These values are computed and plotted against the mean of each predictive parameter. In all the computations, the lead-time of at least 6 h established in §7.7.1 is maintained between the predictive parameters and the surge prediction errors. The bias and confidence intervals are computed with respect to one parameter (univariate) and two parameters (bivariate). In both cases the available data is divided into two: January to June and July to December.

Univariate confidence bounds

Here the confidence intervals are computed using one predictive parameter as a basis. There are roughly 8000 (4000 for each half of the year) samples. The number of bins used is 50 meaning that each bin contains roughly 80 samples. The bias and confidence intervals obtained are shown from Figure 7.40 to Figure 7.44.

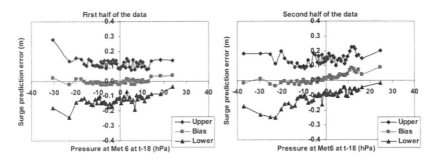

Figure 7.40. Confidence interval with respect to pressure at Met6

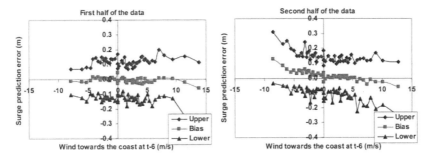

Figure 7.41. Confidence interval with respect to wind speed towards the coast

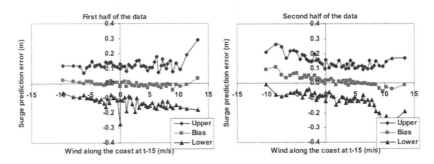

Figure 7.42. Confidence interval with respect to wind speed along the coast

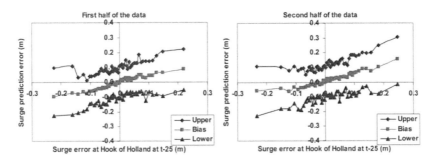

Figure 7.43. Confidence interval with respect to surge errors at t-25 hours

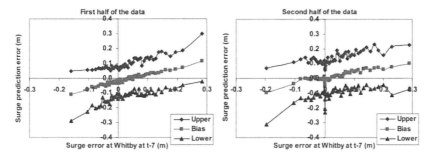

Figure 7.44. Confidence interval with respect to surge errors at Whitby at t-7 hours

The results of the univariate analysis show that there are patterns both in the bias and confidence intervals revealed in the dimension of each selected parameter. In all the cases, the confidence bounds tend to grow wider as the absolute magnitude of the predictive parameter increases. The bias shows a consistent positive slope with respect to past surge errors at Hook of Holland and Whitby. It also shows small negative slopes with respect to the wind components and a small positive slope with the pressure. These are in fact indications of the tendency of the prediction errors. The other observation is that the results obtained from the first and second halves of the data are more or less consistent.

Bivariate confidence bounds

The bivariate confidence intervals are computed using two predictive parameters as a basis, which means that bins are constructed in two dimensions. Since there are roughly 8000 (4000 for each half of the year) samples, only 10 partitions are considered on each predictive parameter meaning that there are 10x10 = 100 bins on the bivariate space with each bin forced to have about 40 samples. The bivariate computations are made using two groups of parameters:

❑ The first group consists of the predicted pressure at Met6 (t-18 hours) and the previous errors at Hook of Holland (t-25 hours).

❑ The second group consists of the observed wind speed along the coast (t-15 hours) and towards the coast (t-6 hours) at Hook of Holland.

The results of the bivariate analysis could not be presented in graphical form but it showed that, in each bin of one predictive parameter, the confidence intervals vary in excess of 10 cm depending on the value of the other parameter, which shows the importance of using multiple predictive parameters. The pattern of the confidence bounds is also consistent between the first and second half of the data.

7.8.2.   Development of the neural network model

Here also a multi-layer perceptron (MLP) type network with the structure shown in Figure 7.45 is used. The input layer of the network is composed of the mean values of each predictive parameter falling in each bin. Thus the number of input nodes is as many as the number of predictive parameters used to compute the bias and confidence intervals. The output layer of the network is composed of three nodes consisting of the bias (mean), and the lower and upper bounds of the 90% confidence interval of the surge prediction errors corresponding to each bin of the predictive parameters.

The advantage of this ANN model over the one used for direct error forecasting (§7.7) is that it helps to forecast not only the bias of the expected surge prediction error but also the range within which the error falls with a 90% confidence level. This helps to account for both the deterministically predictable and unpredictable part of the surge prediction errors, which makes it a more holistic approach to forecasting the surge prediction accuracy of the DCSM model. Its disadvantage is that the number of training examples is limited, consequently limiting the number of parameters used as input in spite of the fact that using all the related parameters as input at the same time gives the ANN more predictive capability. The remedy to this disadvantage is simple - obtain more data.

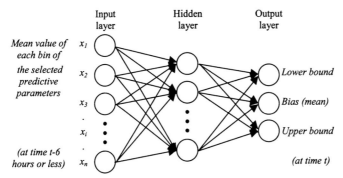

Figure 7.45. Neural network structure for accuracy prediction

The results obtained from the bivariate bias and confidence interval computation are used to develop the neural network. That means that the number of input nodes is two. Also two ANN models are considered with different sets of input data. The first ANN model uses sea surface pressure at Met6 (t-18 hours) and past surge prediction errors at Hook of Holland (t-25 hours) as input. The second ANN model uses the observed wind speed at Hook of Holland towards the coast (t-6 hours) and along the coast (t-15 hours) as input. It is possible to test other combinations of two predictive parameters as input to the ANN; however, with limited number of training examples, there is no compelling reason to justify more tests.

In each case the experiment is conducted by splitting the available data into two: the bias and confidence intervals computed using the data from the second half of 1999 (July-December) are used for training since there are more extreme values in this portion of the data. Those computed from the first half of the data are used to verify the performance of the trained network. Since 100 bins (10 on each of the two parameters) were used to compute the bias and confidence intervals, both the training and testing data consist of 100 examples each. With only 100 examples, two input nodes (mean values of the two predictive parameters) and three output nodes (bias, upper and lower bound of the errors), the number of hidden nodes has to be limited to avoid an under-determined network.

Using pressure and past surge errors as input

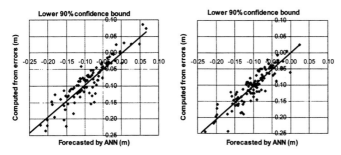

Figure 7.46. Lower bound of 90% confidence (a) Training (b) Verification

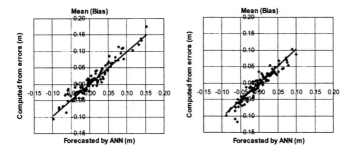

Figure 7.47. Mean (bias) prediction error (a) Training (b) Verification

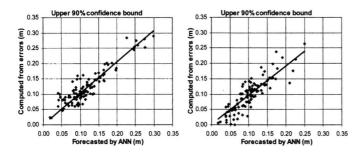

Figure 7.48. Upper bound of 90% confidence (a) Training (b) Verification

Using wind speed along and towards the coast as input

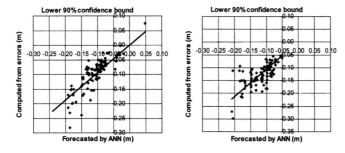

Figure 7.49. Lower bound of 90% confidence (a) Training (b) Verification

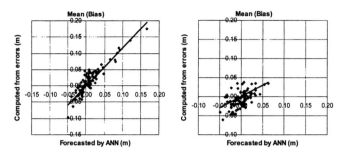

Figure 7.50. Mean (bias) prediction error (a) Training (b) Verification

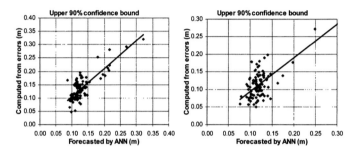

Figure 7.51. Upper bound of 90% confidence (a) Training (b) Verification

Accuracy of the forecasted confidence intervals

The results of training and verification of the ANN accuracy prediction model are shown from Figure 7.46 to Figure 7.51. These scatter diagrams show that the ANN model which uses as input the predicted pressure at Met6 and past surge prediction errors at Hook of Holland performs much better than the one using the wind components. The model that uses the wind components performs better in training than verification. This is also confirmed by looking at the RMSE and the correlation between the computed and ANN forecasted values of the bias and confidence intervals shown in Table 7.8 and Table 7.9.

Table 7.8. Prediction accuracy using pressure and past prediction errors as input

Measure	RMSE (m)		Coefficient of efficiency	
Data	Training	Verification	Training	Verification
Upper 90% bound	0.0243	0.0293	0.817	0.729
Mean (bias)	0.0154	0.0145	0.909	0.894
Lowe 90% bound	0.0264	0.0246	0.853	0.805

Table 7.9. Prediction accuracy using wind along and towards the coast as input

Measure	RMSE (m)		Coefficient of efficiency	
Data	Training	Verification	Training	Verification
Upper 90% bound	0.0282	0.0309	0.615	0.428
Mean (bias)	0.0179	0.0191	0.807	0.099
Lowe 90% bound	0.0310	0.0342	0.715	0.360

7.8.3.   Testing confidence intervals of individual surge predictions

The neural network models used for forecasting the surge prediction accuracy (bias and confidence intervals) are developed and tested using statistically processed data. The input data are the mean of a sample population within a certain range of the predictive parameters. The output data are the computed bias and 90% confidence intervals of the surge prediction errors corresponding to the class of input data.

However, the main purpose of the neural network models is for use in particular surge forecasts, not for statistically computed values. The only way to test whether the ANN model does what it is expected to do is to test it using as input the actual predictive parameters, and to compute the confidence intervals and see if the actual errors from the DCSM fall within the confidence bound. Since 90% confidence intervals are used, only 90% of the surge prediction errors are expected to fall within the forecasted confidence interval.

Out of the two ANN models, the one using surface pressure at Met6 (t-18 hours) and past surge prediction errors at Hook of Holland (t-6 hours) is used to forecast the confidence bounds of the surge prediction at Hook of Holland 6 h in advance. This ANN model is chosen based on the fact that it performs better than the one that uses as input the wind components at Hook of Holland. Since there are no other data to conduct these tests, the verification data (first half of 1999) are used. Hourly data from the second half of alternate months, February, April and June, are used and give the results plotted in Figure 7.52.

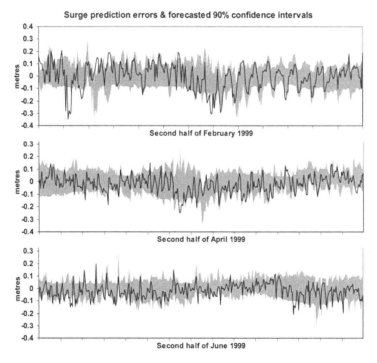

Figure 7.52. Surge prediction errors from DCSM (solid line) and 90% confidence interval forecasted by ANN (shaded area)

The results plotted show that except for a few isolated stretches, most of the error time series falls within the confidence intervals forecasted by the ANN model. Obviously, a visual observation of the time series would not be enough to validate the results and further analysis was carried out to determine what percentage of the surge prediction errors of the DCSM actually fall within the confidence interval forecasted by the ANN model. The results show that 87% of the hourly surge prediction errors in the verification set (January to June 1999) fall within the corresponding confidence intervals forecasted by the ANN model. This is a very good performance compared to the target confidence level of 90% especially taking into consideration the fact that the input data used to train the ANN model are statistical averages whereas those used in this particular test are actual hourly values of the predicted pressure at Met6 and past surge prediction errors at Hook of Holland. Also, it has to be noted that the training and verification data are divided right across the middle of the data for the year 1999, meaning that the presence of seasonal variations between the data of the first and second half of the year cannot be ruled out.

To check the performance of the ANN model further, the absolute magnitude of the difference between the actual surge prediction errors and the bias forecasted by the ANN model are plotted against the corresponding forecasted confidence intervals (which is the difference between the upper and lower bounds) in Figure 7.53. This scatter diagram shows that almost all of the points lie under the diagonal. Some points are above the diagonal, which is because of the choice of a 90% confidence limit.

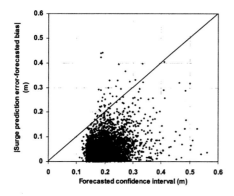

Figure 7.53. Absolute difference between surge prediction errors and forecasted bias versus forecasted confidence intervals

7.9.   Fuzzy rule-based characterization of surge prediction accuracy

In this section an alternative approach of forecasting the surge prediction accuracy is applied. Adaptive fuzzy rule-based modelling is used to characterize the general surge prediction accuracy at Hook of Holland in the form of IF-THEN rules. Again, the errors between model predictions and corresponding historical observations are used as the basis for the analysis. Abebe & Price (2003a) have demonstrated this approach with a water level prediction model of a hypothetical estuary. This case study presents a real life problem to test the applicability of fuzzy rule based modelling in predicting uncertainty.

The fuzzy rule-based system established as a result of such analysis can be applied in real-time surge forecasting since the rules can be formulated offline using the historical performance of the model. A series of such rules covering the input domain of a prediction model help to predict the error even before the corresponding measured values of the surge are available. The method is called adaptive because the fuzzy rules are generated from the data using genetic algorithms.

Data preparation

Like that of the neural network model prediction of surge prediction errors and 90% confidence bounds, the fuzzy rule-based uncertainty characterization model is based on the results obtained from the data relationship analysis. The historical prediction errors from the DCSM model and the selected predictive variables are used. As usual, the approach starts from predetermined and linguistically meaningful fuzzy sets on the output as well as the input domain.

The input data to the fuzzy model are two selected predictive parameters whereas the output is the surge prediction error of the DCSM at Hook of Holland. The next step is to cover the range of input and output data by meaningful fuzzy sets. The minima and maxima of the predictive parameters and the target data used to train the rules are shown in Table 7.10.

Table 7.10. Minima and maxima of selected predictive parameters

Data	Minimum	Maximum
Surge error at Hook of Holland	-0.45	0.45
Surge error at Whitby	-0.50	0.50
Wind along the coast at Hook of Holland	-13.42	18.13
Wind towards the coast at Hook of Holland	-10.93	17.44
Pressure at Met6	-50.00	30.00

For each of the predictive parameters used as input to the fuzzy rule-based system, five fuzzy sets and their corresponding membership functions are constructed. The fuzzy sets range from –HIGH (negative high) to +HIGH (positive high). For the surge prediction errors, negative and positive represent the algebraic sign of the error. For the pressure, these signs represent whether it is below or above standard atmospheric pressure. For the wind components, negative and positive represent the direction of the wind. Southwesterly winds are positive along the coast and northwesterly winds are positive towards the coast.

Triangular membership functions are used for all the variables. Triangular membership functions are defined using three values: the minimum and maximum values having a membership of zero and a middle value having a membership of one. Consequently, the membership functions for surge prediction errors, wind components and sea surface pressure are constructed as shown in Figure 7.54, Figure 7.55 and Figure 7.56, respectively.

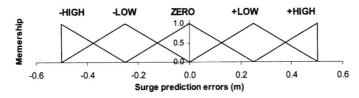

Figure 7.54. Membership functions on surge prediction errors

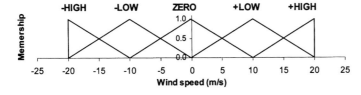

Figure 7.55. Membership functions on wind components

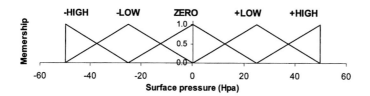

Figure 7.56. Membership functions on surface pressure

Generating the rules

Construction of the membership function completes the IF part and the THEN part of the rules. The next step is to connect them. Each connection is a fuzzy rule relating the input and output fuzzy sets in IF-THEN form. The process of connection is posed as a combinatorial optimisation problem using genetic algorithms. The payoff function is the inverse of the RMSE between the output of the FRBM and corresponding training data. Posing the problem as such ensures that the results are linguistically sound.

The number of inputs used to generate the rules is limited to two:

❑   The first set of input data are the past surge prediction error at Hook of Holland (t-25 hours) and the sea surface pressure at Met6 (t-18 hours).

❑   The second set of input data are composed of the wind components along the coast (t-15 hours) and towards the coast (t-6 hours) at Hook of Holland.

It is logical to assume that the use of more than two input data at once, say four of the above parameters, would result in more representative fuzzy rules. However, this has been avoided intentionally for two reasons: (1) the amount data available to formulate the rules is limited to one year and is not sufficient to train the number of rules that grows geometrically with the number of inputs, and (2) the resulting system of rules would be more human understandable if they are presented in a lower dimensional matrix such as in one or two dimensions.

Results

The generated fuzzy rules are shown in Figure 7.57. The rules generated using pressure at Met6 and past surge prediction errors at Hook of Holland (Figure 7.57(a)) are in accordance with the pattern obtained from the data relationship analysis and the univariate and bivariate confidence interval computations. However, the rules generated using the wind components at Hook of Holland as input (Figure 7.57(b)) do not show much variation in the rule outcomes. This corresponds to the fact that the bias does not show a dramatic change in magnitude with respect to the wind components. This does not falsify the rules obtained; rather it shows that it is this much information that the wind components can tell in the form of fuzzy rules.

(a)

SURFACE PRESSURE AT MET6 (t-18 hours)	SURGE ERROR AT HOOK OF HOLLAND (t-25 hours)				
	-HIGH	-LOW	ZERO	+LOW	+HIGH
-HIGH	-HIGH	-LOW	ZERO	ZERO	+LOW
-LOW	-LOW	-LOW	ZERO	ZERO	+LOW
ZERO	-LOW	ZERO	ZERO	ZERO	+LOW
+LOW	-LOW	ZERO	ZERO	+LOW	+HIGH
+HIGH	-LOW	ZERO	+LOW	+LOW	+HIGH

(b)

WIND TOWARD THE COAST AT HOOK OF HOLLAND (t-6 hours)	WIND ALONG THE COAST AT HOOK OF HOLLAND (t-15 hours)				
	-HIGH	-LOW	ZERO	+LOW	+HIGH
-HIGH	+LOW	+LOW	+LOW	ZERO	ZERO
-LOW	+LOW	+LOW	ZERO	ZERO	ZERO
ZERO	+LOW	ZERO	ZERO	ZERO	ZERO
+LOW	ZERO	ZERO	ZERO	ZERO	ZERO
+HIGH	ZERO	ZERO	ZERO	-LOW	-LOW

Figure 7.57. Fuzzy rule based characterization of surge prediction accuracy in terms of (a) past surge errors and pressure at Met6 (b) wind components along and towards the coast

The way a fuzzy inference engine applies the rule system is as follows. For example, let the predicted pressure at Met6 18 h before the expected surge be −25 hPa whereas the error in predicted surge at Hook of Holland 25 h before the expected surge be +0.375 cm. The pressure belongs to the fuzzy set -LOW with membership of 1 (Figure 7.56). The past surge prediction error belongs to +LOW with membership of 0.5 and +HIGH with a membership of 0.5 (Figure 7.54).

Two of the rules shown in Figure 7.57(a) will be activated:

1. IF Pmet6 is −LOW and PastError is +LOW then ExpectedError is ZERO

2. IF Pmet6 is −LOW and PastError is +HIGH then ExpectedError is +LOW

The expected surge prediction error is therefore between ZERO and +LOW which indicates a slight overestimation of the surge.

Assuming the degree of fulfilment (DOF) of a rule to be the product of the membership of its premises (product inference), the first rule has a DOF=1*0.5=0.5 which in this case is the same as the DOF of the second rule.

To obtain a crisp output from the rule system (or defuzzify), the DOFs of the rules can be used as weights to determine the contribution of each rule. The crisp outcome is the centroid of 0.5(ZERO) + 0.5(+LOW) of the fuzzy triangles shown in Figure 7.54, which becomes 0.125 m.

The way it is applied here, the fuzzy rule system is optimised for linguistic rule extraction and not for numerical precision. Therefore, the generated fuzzy rules can be used to obtain an impression of the expected surge prediction accuracy in linguistic terms, for example, whether the surge is over- or under-estimated and to what degree, before the event is observed. If it is intended to apply this technique to obtain numerical outputs, it is recommended to use other algorithms to train the rule system.

7.10. Conclusions and discussion

The problem of forecasting the accuracy of surge forecasts on the Dutch coast, particularly at Hook of Holland, has been approached step by step from a general analysis of the relationship between several variables to a more detailed study focusing on selected variables using state-of-the-art techniques. The data relationship analysis has revealed the relative importance and type of information available in several variables. These findings are already summarized in §7.6.

The tests conducted on direct forecasting of surge prediction errors using neural network models showed the relative importance of each of the selected parameters. The neural network model managed to forecast the surge prediction errors to some degree. However, the results obtained were not adequate to use error forecasting as a sole means of managing the surge prediction accuracy. A considerable portion of errors and consequent uncertainty remained in the surge prediction. This gave rise to the search for an alternative approach to make a more complete assessment of the surge prediction accuracy.

Consequently, the bias and confidence intervals of the surge prediction errors were computed with respect to the selected predictive parameters. The confidence intervals as well as the bias showed patterns that depend on the magnitude of the predictive parameters. The tests repeated on a bivariate basis showed that even when the magnitude of one predictive parameter is fixed, the bias and confidence intervals show a considerable variation with respect to the other predictive parameter, which is an indication that, if more data are available, the use of more predictive parameters together can improve the results.

The results of the bivariate computations were used to train and verify a neural network model that targets not the errors directly but the bias and confidence intervals. The network performed very well when the predicted pressure and past surge errors at Hook of Holland are used as input compared to using the wind components as input. This does not mean the wind components are not related; rather it highlights the dominance of the tide-induced periodic pattern in the surge prediction errors. The neural network that uses past surge prediction errors and predicted pressure has an obvious advantage in that it has input data composed of the historical errors, which accounts for the periodicity, and the pressure, which accounts for meteorological factors.

The ANN that uses the past surge prediction errors at Hook of Holland and the predicted pressure at Met6 was applied to forecast the 90% confidence intervals for individual hourly surge predictions at Hook of Holland. The results showed that the forecasted confidence intervals contained 87% (instead of 90%) of the predicted surges, which is a very outstanding performance taking into consideration that a neural network model trained with statistically computed values is applied to forecast confidence intervals for particular hourly surges. Even though this ANN is trained considering a 6 h horizon, it can be applied to forecast the bias and confidence bounds up to a horizon of 18 h with the same accuracy. This is because its input data are the predicted pressure at t-18 hours and previous surge prediction errors at Hook of Holland at t-25 hours.

The fuzzy rule-based modelling approach resulted in a set of rules that helped to characterize the surge prediction accuracy with an array of overlapping rules. More descriptive rules were obtained when the predicted pressure and previous surge prediction errors were used as input. The rule system can be used as a look-up chart of the expected surge prediction accuracy in the form of human language. By feeding the actual magnitude of the input data (for example the predicted pressure at Met6 and the previous surge prediction errors at Hook of Holland) the rule system can be evaluated and defuzzified to obtain the expected bias. However, if the intention is to forecast the expected bias of surge prediction errors in a numerical form using a FRBM is intended, it is recommended to use other rule training algorithms that give more importance to numerical accuracy rather than a linguistic interpretation of the generated rules. The fuzzy rule-based approach used only two selected predictive parameters at a time. Theoretically, it is possible to use more predictive parameters as input. However, the number of rules to be trained increases geometrically with the number of input parameters and, consequently, so does the number of training examples needed. The main drawback of using multiple dimensions is that the generated rule system can be less comprehensible.

This particular application problem gave additional dimensions to the applicability of the complementary modelling approach. It demonstrated that it is not only prediction and forecasting of errors that complementary models can do. Rather, complementary models can be trained to forecast the accuracy of physically based models. The accuracy can be defined in the form of bias and confidence intervals and in the form of linguistic descriptions of the expected errors. Even though ANN and FRBM techniques were used in this particular application, there is no reason why any data-driven modelling technique appropriate to the problem cannot be used in complementary modelling. With the use of more data, both the ANN and FRBM techniques can be refined further. Splitting one-year data for training and verification might be affected by seasonal variations and cause a difference between the training and testing patterns. Both techniques are very promising and provide different types of information regarding the surge prediction accuracy.

More work can be done in this respect, such as:

❑ Carrying out the whole study for a larger set of data. This study has been limited to the year 1999. The data had to be split in the middle to train and verify the data-driven models. If there are considerable seasonal variations in the data, the performance of the data-driven models could be affected. It is therefore worth testing the methods with observed data and DCSM predictions for other years.

❑ Testing information flow for more specific aspects of the data. In the present study, data relationship analyses are carried out for the whole data. With the presence of more data, data relationship analysis, for example, on stormy condition or specific wind directions, could provide more insight.

❑ Confirming the applicability of the methodology to longer forecast horizons.

❑ Applying the methodology to radically different situations such as flood generation in river basins.

The results of the data relationship analysis give insight to the possibilities of improving the DCSM itself. First let us consider the dominant periodic pattern in the surge prediction errors that corresponds with the tidal cycle. Surge is not a periodic process but could have periodic components since the meteorological forces act on the sea at different tidal states of the sea which introduces periodic patterns. However, surge prediction errors should not have any periodic patterns unless the non-linear relationship between surge and tide is not properly represented in the model. This suggests that it is necessary to sort out the relationship between surge and tide in a better way than it is done at present. In an ideal situation, there should not be any periodicity in the surge prediction errors.

Another important indication is the fact that the meteorological variables are related to the surge prediction errors. Again, meteorological forces (wind and pressure) are physically related to the surge. If the model properly represents this relationship, there should not be any relationship between the meteorological forcings and the surge prediction errors. This suggests that the way the surge is computed using the meteorological forcings needs to be re-evaluated.

No distinct relationship was found between wind prediction errors and surge prediction errors. However, it is also obvious that improving the quality of the meteorological forecasts will improve the surge prediction accuracy.

PART IV. EVALUATION

CHAPTER 8. CONCLUSIONS, DISCUSSION AND FUTURE WORK

8.1. Conclusions

The research carried out in this study has indicated that information theory-based techniques help to determine how errors in the results of physically based computational models are related to the input and output data for the models. In particular the techniques enable information to be obtained about which data can be used to assist in the recovery of the errors, and to generate insight into the time dynamics inherent in the relationship between the data and model errors. It has been shown that errors from several types of computational models share varying degrees of information with their respective input data, output data and the state variables. This insight has helped to identify particular time series that share a maximum amount of information with the residual errors. In turn, these time series are used subsequently to construct a complementary data-driven model, which can be used to forecast the expected errors of the model. In particular, the average mutual information (AMI) analysis has helped not only in the development of data-driven models but also in determining parameters for physically based computational models such as the wave speed of flow routing models. In addition AMI has facilitated an analysis of the time dynamics of the storm surge observed at several locations in the North Sea.

Referring to the case studies done on the application of complementary models, the methodology has helped to improve the prediction accuracy of models having different sources of uncertainty. For the hypothetical estuary model, the problem is related to the numerical solution. In the models of the Meuse and Wye rivers, the main source of uncertainty is the lateral inflow along the river reaches. For the model of the Rhine, the primary sources of uncertainty are the flow distribution into the branches, the operation of the gates and the effect of tides. For the North Sea model (DCSM), the suspected sources of uncertainty are the relationship between the storm surge and the tide, the accuracy of meteorological forcings and their relationship to the surge. In all these cases, the results indicated that forecasting the errors of the physically based computational model with a complementary data-driven model considerably improves the forecast accuracy. This is demonstrated by the consistent improvement in the performance indices (such as root mean square, mean absolute error, and coefficient of efficiency) of the predictions.

The case study made with the DCSM was different from the other case studies in that direct forecasting of errors with a complementary model failed to improve surge predictions significantly. Therefore the problem was reformulated such that the complementary model was trained to forecast the accuracy of physically based model and not the errors directly. The accuracy was defined using the bias and 90% confidence intervals as well as the linguistic descriptions of the expected errors. This demonstrated that a complementary model does not only provide a forecasting of the errors: it can also be trained to forecast the accuracy defined in some other form.

With the use of fuzzy rule-based models it has been demonstrated that the expected prediction accuracy of a model can be characterized by a system of linguistically described rules that relate selected state variables to the model prediction accuracy. Two case studies involving hydrodynamic models are explored; in both cases applying genetic algorithm on the historical data generates the rules. The generated rule system can be used as a look-up chart of the expected accuracy of the model prediction.

The case studies have also indicated that, along with appropriate analysis techniques, the model errors can be used as indicators to make further improvements to the physically based computational models. This is because the patterns in the errors and the varying level of information they share with the state variables can be used to characterize deficiencies in a physically based model. For example, in Chapter 7, it has been suggested that, based on the dominant periodic pattern in the storm surge prediction errors and its coincidence with the tidal cycle, it is necessary to sort out the non-linear relationship between surge and tide in a better way than is done at present. This is because, in an ideal situation, there should not be any periodicity in the surge prediction errors because the surge is not a periodic process. Recommendations for improvement of models are given in other case studies as well. However, in order to unveil patterns in prediction errors of models and to trace them back to the model, one should know what to look for and where to look for it.

The thesis has demonstrated that complementary modelling is much more than simply predicting the errors of simulation models. It is about considering the gap between a mathematical model and the physical processes it purports to represent, and identifying the gap as a separate process in its own right. In this way a specific model can be developed for the errors defining the gap. It is one way of making sense out of the historical errors of a model. The approach opens up a range of possibilities from improving model predictions to obtaining valuable information in order to help understand the physical processes as well as the behaviour of the simulation model. The historical errors of the simulation model are treated as any other data 'observed' from this other 'process'. Complementary modelling enables a stereoscopic view of the physical processes in the sense that it allows the modeller to see the physical domain through two different modelling paradigms working in an optimal manner. A successful application of complementary modelling essentially involves an understanding of the principles on the basis of which both data-driven models and physically based models represent a particular physical system.

8.2. Discussion

Type of data-driven models that can be used: Most of the case studies in this thesis make use of ANN and fuzzy rule-based models as complementary models. However, in principle, any data-driven modelling approach can be used depending on its suitability for the particular problem with regard to the type of the process and the availability of data. In the case study on the Rhine model, a linear regression model was used for the complementary model, which happened to be a good choice. There are many other data-driven modelling techniques such as nearest neighbourhood, local linear models based on chaos theory, genetic programming, support vector machines and model trees. In applications where the complementary model is intended to serve as a means of classifying the expected accuracy, classification techniques such as Kohonen networks, clustering algorithms and fuzzy rule-based models can be applied. There is now a wealth of literature in all of these modelling techniques.

Physically based computational models to which complementary modelling can be applied: It has been demonstrated that complementary modelling can be applied successfully to physically based computational models for the hypothetical estuary and the North Sea. It has also been applied to a conceptual rainfall-runoff model of the Sieve Basin, Italy. Therefore, it has been shown to be applicable to both physically based and conceptual computational models.

Complementary modelling can be applied to models that are already in practical use as demonstrated with the DCSM model. However, it can also be applied to models that are under development. The Rhine and Meuse forecast models were developed as a part of this study. The advantage of using complementary modelling in the case of models under development is that the complementary model can be integrated easily. However, the prospective use of complementary modelling should not undermine the effort that the modeller puts towards developing the physically based computational model. Also, if changes are made to the primary model (to its structure or parameters), then its error pattern might change and the complementary model will have to be re-trained to adapt to the error patterns of the altered primary model.

Model outputs that can be updated by a complementary model: It has been demonstrated that both intermediate and final outputs can be updated using a complementary model. For instance, in the Rhine model, the final outputs at Hagestein are updated whereas for the Meuse model, intermediate outputs at Venlo are updated. However, in order to update intermediate outputs the primary model, it has to be possible to integrate the complementary model at intermediate stages of the primary model, which may be difficult for models that are already in use.

Developing a complementary model involves errors at target locations where outputs are to be updated, which implies that observed data have to be available at these locations. This is a disadvantage in that it is not possible to update predictions at locations where observed data are not available.

Data needs: The complementary model can use any relevant data so long as the data help in forecasting the expected errors or prediction accuracy. This even includes data that are not used by the primary model. However, these data have to be available in operational forecast settings.

Computational demand: Stochastic uncertainty management techniques generally require a substantial number of model simulations since they depend on multiple model runs with generated parameters to approximate closely the *a priori* probability distribution of uncertain parameters. This affects the applicability of the management techniques in models that demand significant computer time particularly in real-time forecasting applications. However, the task that demands significant computer time in complementary modelling is the development of the data-driven model, which is done offline. Once it is developed, executing an application, such as a neural network model, takes only a fraction of a second, which implies it is not an operational issue. The procedure does not involve any additional runs of the primary model that would otherwise be necessary without a complementary model.

8.3.   Contribution

This thesis has made a number of important contributions to modelling real-world water based systems.

8.3.1.   Application of information measures

The study has demonstrated various ways in which information theory-based techniques can be applied in the process of developing physically based process models as well as complementary data-driven models. In particular, the applicability of the average mutual information (AMI) measure in the determination of parameters for models and in the analysis of the time dynamics of observed and error time series has been demonstrated.

The weighted mutual information (WMI) has been developed on the basis of the average mutual information (AMI) as a measure of biased information. The AMI measures the 'average' information shared between two data series. When it is applied with varying lag times between the two data series, the lag time corresponding to the maximum AMI is an indication of the separation time between the two data. However, the temporal separation between the two data series can vary depending on the magnitude of data in either set.

The WMI measure has more enhanced features that enable it to determine the separation time for particular ranges of the data. This makes WMI more suitable to identify the time dynamics between two data series. When two data series are related with each other at different lag times depending on the magnitude of one of the data sets, WMI can be used to mark the separation line between the shifts in the time dynamics of the relationship. This feature can be used in selecting input-output relationships for data-driven models.

In §3.8, the application of the WMI has been demonstrated with a case study establishing the relationship between the wave speed and the discharge in rivers by calculating the time at which flood waves propagate from upstream to downstream of a river. Its application is particularly important when the only available data is the discharge or water level time series at two ends of a river reach.

The WMI has plenty of other potential application areas in understanding the time dynamics of processes. For instance, it can be used to analyse the response time of coastal surges for particular storms. Unfortunately, it was not possible to explore this application in Chapter 7 because only one year of data was available for the study.

8.3.2.   Complementary modelling

The study has demonstrated the extensive application of the complementary modelling procedure with a variety of case problems involving different types of physical processes and models. It enables an optimal use of physical insight about the processes and the historical data. Its application in real-time forecasting is demonstrated with several case studies. Since the data-driven model works at the 'end of the line', the procedure is computationally efficient in that it does not involve any additional runs of the physically based computational model. This also makes it equally applicable to models that are already in use as well as to those under development. The important contribution in this respect is the possibility of forecasting the accuracy of forecasts made by physically based models in the form of bias and confidence bounds with the use of data-driven models. This has a lot of potential applications in that it enables the development of a separate model to determine the forecast accuracy of physically based forecast models.

8.3.3.   Fuzzy characterization of model prediction accuracy

Combining the complementary modelling concept and the fuzzy rule-based modelling technique enables a new form of characterizing the overall prediction accuracy of models. The prediction accuracy of a model is presented as a system of fuzzy IF-THEN rules that relate selected state variables to the expected errors. One advantage of this approach is that it deals with the overall prediction accuracy because it works on the gap between the model prediction and corresponding observation. The other advantage is that the analysis results are presented in a linguistic form that is easily understandable. This technique can be used to anticipate the accuracy of models in real-time forecasting.

8.3.4.   Learning from errors

Finally, the study has presented a new way of looking at model errors, that is, the 'gap' between model predictions and corresponding observations. All the case studies involving complementary modelling have something to do with model errors. Errors contain valuable information that can be used in various ways. In particular, the information can be used to set up a model to predict errors and minimize the prediction uncertainty.

It is also possible to obtain feedback about the physically based computational model. In the applications demonstrated in Chapters 6 and 7, it is shown that with a combined use of physical insight and appropriate analysis techniques, patterns in the errors of a physically based model can be traced back to physical processes that have been inadequately formulated in the physically based model. This information can be used to refine the physically based model.

8.4.   Possibilities for further research

This thesis has only begun to address some of the interesting possibilities for complementary modelling.

8.4.1.   Application to other water-based models

In this thesis, complementary modelling is applied to rainfall-runoff, river flow routing and coastal hydrodynamic models. The methodology has considerable potential in application to radically different situations such as flood generation in river basins or storm surge prediction in coastal waters. It is also important to investigate its applicability to water quality models. Often, water quality simulations are carried out with the flow velocities predicted separately by hydrodynamic computations. Therefore, improved hydrodynamic computations may well help to improve water quality simulations. However, it would also be good to test the applicability of complementary modelling to water quality models directly in that sizeable uncertainties exist in the predictions of such models. Similar arguments support the application of complementary modelling to the vexed problem of sediment transport, though as with water quality modelling the success of such an application is highly dependent on the availability of an adequate quantity of good quality data. Another potential area of application is to ground water modelling. A potential problem in application to ground water models however is that there are generally less observed time series data for ground water studies than for surface water.

8.4.2.   Spatial extension of complementary modelling

Physically based computational models that involve spatial grids are generally used to make predictions at multiple locations. River, estuarine, and oceanic models can be taken as examples. For some of the locations where predictions are made, observed data (such as discharge, water level, current) are available. For most locations, however, there is no observed data. According to the way it is formulated in this thesis, the complementary modelling approach is applied to update model outputs at locations where observed data are available. This is because the methodology is based on the difference between model predictions and observed data.

In this respect an important but challenging area for future research is to investigate the possibility of applying updates at locations where observed data are not available. This is potentially useful in practical applications such as flood inundation and ground water models. For example, updating water levels at few locations will not be adequate to generate a more accurate flood inundation map.

8.4.3.   Symbolic interpretation of the complementary model

The complementary model is defined as a data-driven model of the gap between the primary (physically based) model and the actual process. It is possible that complementary model contains information that can help in refining the primary model. This potentially can lead to improvements in the primary model and to a better understanding of the physical processes. It would be challenging undertaking to study the structure of the complementary model and obtain valuable information that can be traced back to defects in the physically based model.

For instance, if the complementary model is an artificial neural network, it will be interesting to extract some type of symbolic information from the network structure and the corresponding patterns in the weights in the form of, say, a partial differential equation that can be related to the primary model.

8.4.4. Embedding data-driven models within physically based models

The complementary modelling procedure presented in this thesis follows a loose connection between the data-driven and physically based models in that they are parallel but the physically based model is practically independent. It is yet another possibility for further research to explore a tighter embedding of the data-driven model within a physically based model. Such an approach corresponds to a more direct form of data assimilation for physically based computational models than using techniques such as Kalman filters. The data-driven model can be formulated to accomplish specific tasks that are not physically well defined. For instance, the conveyance function in de Saint Venant flow equations has several alternative definitions since it is not physically well defined. There may be a possibility to use a data-driven model in the structuring of the conveyance function. Another possibility is to embed a data-driven model in a numerical model at a grid cell level. If a data-driven model can learn the relationship between the state variables in a computational grid cell, it might help overcome some of the more computationally demanding routines within numerical models.

Samenvatting

Informatie Theorie en Kunstmatige Intelligentie om Onzekerheden in Hydrodynamische en Hydrologische modellen te beheren

Mathematische modellen van waterhuishouding systemen zijn in de afgelopen tiental jaren veelvuldig toegepast. De meeste modellen zijn gebaseerd op mathematische formulering van physische wetten zoals de wet van behoud van massa en momentum. Dit soort modellen zijn breed toegepast bij voorspellingen en verwachtingen. Alhoewel een ander soort modellen steeds meer terrein wint, dit zijn gegevens-gestuurde modellen, zoals kunstmatige neurale netwerken, genetisch programmeren, fuzzy logic generatoren, support vector machines, etc. Hierin wordt het model gebaseerd op alleen historische gegevens in plaats van de physische eigenschappen van de specifieke toepassing. Met de toename in de toepassing van physisch gebaseerde modellen wordt de onzekerheid steeds meer belangrijk omdat o.a. wat het model voorspelt en wat er wordt gemeten van het physisch systeem niet perfect past. Geen enkel mathematisch model kan het physische systeem perfect simuleren. Dit onderzoek probeert het verschil tussen de physisch gebaseerde modellen en de observaties bij het gebruik van gegevens-gestuurde modellen en informatie gebaseerde principelen te minimaliseren.

Dit onderzoek is gebaseerd op het feit dat in de ontwikkeling van physisch gebaseerde en gegevens-gestuurde modellen er veel verschillende gegevens nodig zijn, voornamelijk physisch inzicht voor de physisch gebaseerde modellen en physische gegevens voor de gegevens gestuurde modellen zijn belangrijk, wat er op duidt dat ze complementair zijn en daarom optimal toegepast. Dit leidt tot twee basis veronderstellingen. De eerste veronderstelling is dat de voorspellingen van een physisch gebaseerd model heel erg verbeterd kan worden door het gebruik van een gegevens-gestuurd model van de residuals en daardoor verminderd de onzekerheden systematisch. De tweede veronderstelling gaat over hoe ver je kan gaan met het verbeteren van het physisch gebaseerd model. Het toegevoegde gegevens-gestuurde model heft een deel van de resterende fout op tussen het physisch gebaseerde model en de waarnemingen. Dit gebeurt doordat de resterende fout gerelateerd wordt met geselecteerde variabelen. Het gedeelte van de resterende fout dat met het gegevens-gestuurde model wordt verklaard is daardoor een indirecte waarneming van het aanwezige potentieel om physisch gebaseerde modellen te verbeteren.

Het tweede gedeelte van de thesis, hoofdstuk 3, 4 en 5, beschrijven de ontwikkelde en toegepaste methodologie. In het begin zijn theoretisch gebaseerde principes bestudeerd in relatie tot onzekerheden. De gemiddelde overeenkomstige informatie meting is een belangrijk instrument in het onderzoek, omdat het gebruikt kan worden om te evalueren hoe veel je te weten kan komen over een gegevens set als een andere gegevens set al bekend is. De gemiddeld overeenkomstige informatie waarneming is verder ontwikkeld tot een gewogen overeenkomstige informatie in een bepaalde reikwijdte van de gegevens mogelijk maakt. De toepassing wordt gedemonstreerd door het probleem van het berekenen van vloedgolfsnelheden met gebruik van afvoer hydrografen in rivieren. In hoofdstuk 4 worden de basis principes van kunstmatige neurale netwerken en fuzzy set theorie bestudeerd. Deze technieken worden gebruikt om intelligente gegevens-gestuurde modellen te ontwikkelen in de meeste case-studies in deze thesis. In relatie tot de principles van onzekerheden invariantie geven de fuzzy set theorie en de Monte Carlo simulatie gelijksoortige informatie met betrekking tot de onzekereheden in een groundwater verontreinigings transport model met onzekerheids parameters. Hoofdstuk 5 introduceert het concept van toegevoegd modelleren dat een brug slaat tussen een physisch systeem en zijn (physisch gebaseerd) model door middel van het toevoegen van een gegevens-gestuurd model. Een toegevoegd model kan worden gebruikt om de fouten in de voorspelling van een physisch gebaseerd model te voorspellen en vervolgens upgedate voorspellingen te genereren en de verwachte zekerheden te voorspellen in de vorm van zekerheidsgrenzen en linguistische metingen die gebruikt kunnen worden als mate van onzekerheid. De methodologie is getest met een hydrodynamisch model van een hypothetisch estuatium, een rivier overstromings routing model en een conceptueel regenval - afvoer model.

De toepassingen in de praktijk van toegevoegde modelleren met bestaande operationele modellen en modellen ontwikkeld gedurende deze studie worden gedemonstreerd in het derde gedeelte van deze thesis. De eerste toepassing is bij afvoermodellen voor de Rijn en de Maas in Nederland. Voor beide rivieren geldt dat het toegevoegd modelleren heeft bijgedragen in het verbeteren van het voorspellen van afvoeren in vergelijking met de voorspellingen van de neurale netwerk modellen en physisch gebaseerde modellen apart. Het heeft ook geholpen om processen die niet in de originele modellen zitten toe te voegen zoals getij effecten in het Rijn model en laterale stroming in het Maas model.

De tweede toepassing heeft te maken met het Dutch Continental Shelf Model (DCSM), een 2D model dat gebruikt wordt om water niveaus, opstuwing en snelheden te voorspellen. Deze studie verschilde van andere case studies doordat het direct voorspellen van fouten met het toegevoegde model de stromings voorspellingen niet significant verbeterde. Het probleem was daarom herformuleerd zodat het toegevoegd model de nauwkeurigheid van de physisch gebaseerde modellen zou bepalen in plaats van direct de fouten. De nauwkeurigheid is bepaald met een bias en 90% zekerheidsinterval en een linguistische beschrijving van de verwachte fouten. Dit liet zien dat een toegevoegd model niet alleen de fouten voorspeld, maar het kan ook worden gebruikt om de nauwkeurigheid in een andere vorm te bepalen. In dezelfde case-studie wordt de toepassing van fuzzy-regel gebaseerde modellen gebruikt om de nauwkeurigheid van stromings voorspellingen te demonstreren in de vorm van IF-THEN regels. De regels relateren de bekende zee status of meteorologische variabelen aan de verwacht nauwkeurigheid van de stromingsvoorspellingen in een linguistische vorm.

Toegevoegd aan de mogelijkheden van voorspellen van de fout en nauwkeurigheid van de model voorspellingen, hebben de studie geconstateerd dat samen met toepasselijke analyse technieken, indicatoren voor verdere verbeteringen bij physisch gebaseerde modellen. Bijvoorbeeld, de fouten in storm opstuwing voorspellingen van de DCSM betstaat uit een dominant periodiek patron en de periodiciteit valt samen met de getij cyclus. Dit bevestigd dat het nodig is om de niet-lineaire relatie tussen de opstuwing en het getij beter te beschrijven dan die nu is gedaan. In een ideale situatie zou er geen periodicteit zitten in de opstuwing voorspelling fouten omdat een opstuwing geen periodiek process is. Aanbevelingen voor het verbeteren van andere case-studies zijn ook gegeven.

Het onderzoek heeft bevestigd dat naast zijn toepassing zonder het draaien van het primair (physisch gebaseerd) model te beinvloeden toegevoegde modellen een scala aan andere voordelen met zich meebrengt zoals de mogeljkheid van het toevoegen van gegevens die niet door het primair model worden gebruikt en toch niet meer computertijd vergt. De mogelijkheid om informatie te krijgen dat helpt in de verbetering van de physisch gebaseerde modellen is gedemonstreerd, speciaal voor de DCSM case studie. De toegevoegd modeleer methodologie is ontwikkeld zodat het niet specifiek toegepast kan worden in een model in een bepaalde situatie. Het kan worden toegepast in hele verschillende modellen van physische systemen naast degene die onderzocht zijn in deze thesis.

References

Abarbanel, H. 1996. *Analysis of observed chaotic data*. Springer, New York.

Abbott, M.B. 1966. *An introduction to the method of characteristics*. Thames and Hudson, London.

Abbott, M.B. 1979. *Computational Hydraulics; Elements of the theory of free surface flows*. Pitman Publishing ltd., London.

Abbott, M.B. 1991. *Hydroinformatics: information technology and the aquatic environment*. Avebury Technical, Aldershot, UK.

Abbott, M.B., Bathurst, J.C., Cunge, J.A., O'Connell, P.E. & Rasmussen, J. 1986a. An introduction to the European Hydrological System - Systeme Hydrologique Europeen, "SHE", 1: history and philosophy of a physically-based, distributed modelling system, *J. of Hydrology*, 17, 45-59.

Abbott, M.B., Bathurst, J.C., Cunge, J.A., O'Connell, P.E. & Rasmussen, J. 1986b. An introduction to the European Hydrological System - Systeme Hydrologique Europeen, "SHE", 2: structure of a physically-based, distributed modelling system, *J. of Hydrology*, 17, 61-77.

Abebe, A.J. & Price, R.K. 2000. Application of neural networks to complement physically based hydrodynamic models, *Hydroinformatics 2000 CDRom of Proc.*, 23-27 July 2000, Iowa Institute of Hydraulic Research, Iowa City, USA.

Abebe, A.J., Guinot, V. & Solomatine, D.P. 2000a. Fuzzy alpha-cut vs. Monte Carlo techniques in assessing uncertainty in model parameters, *Hydroinformatics 2000 CDRom of Proc.*, 23-27 July 2000, Iowa Institute of Hydraulic Research, Iowa City, USA.

Abebe, A.J., Solomatine, D.P. & Venneker, R.G.W. 2000b. Application of adaptive fuzzy rule-based models for reconstruction of missing precipitation events, *Hydrol. Sc. Journal*, 45 (3) 425-436.

Abebe, A.J. & Price, R.K. 2002a. Complementary models in managing model uncertainty, *Proc. Hydroinformatics 2002*, 1-5 July 2002, Cardiff, UK, IWA Publishing, 2, 1382-1387.

Abebe, A.J. & Price, R.K. 2002b. Enhancing flood forecasts via complementary modelling, *Proc. 2nd Int. Symposium on Flood Defence*, 10-13 September 2002, Beijing, China, Science Press, New York, 2, 853-860.

Abebe, A.J. & Price, R.K. 2002c. Applying information theoretic approaches in flood propagation, *Proc. 2nd Int. Symposium on Flood Defence*, 10-13 September 2002, Beijing, China, Science Press, New York, 2, 1200-1206.

Abebe, A.J. & Price, R.K. 2003a. Characterization of prediction uncertainty using an adaptive fuzzy rule based technique, *Modelling and Simulation 2003*, 366-371, 24-26 February, Palm Springs, California.

Abebe, A.J. & Price, R.K. 2003b. Managing uncertainty in hydrological models using complementary models, *Hydrol. Sc. J.* 48 (5) 679–692.

Abebe, A.J., Price, R.K., Dillingh, D. & Verlaan, M. 2003. Forecasting flows on the River Meuse, *Proc. XXX IAHR Congress*, 24-29 August 2003, Thessaloniki, Greece.

Abebe, A.J. & Price, R.K. 2004. Information theory and neural networks for managing model uncertainty in flood routing, *J. of Computing in Civil Engineering*, ASCE (*in press*).

Akaike, H. 1974. A new look at the statistical model identification, *IEEE Transactions on Automatic Control*, 19 (6) 716-723.

Babovic, V., Canizares, R., Jensen, H.R. & Klinting, A. 2001. Neural networks as routine for error updating of numerical models, *J. Hydraulic Engineering*, ASCE, 127 (3) 181-193.

Babovic, V., Keijzer, M. & Bundzel, M. 2000. From global to local modelling: a case study in error correction of deterministic models, *Hydroinformatics 2000 CDRom of Proc.*, 23-27 July 2000, Iowa Institute of Hydraulic Research, Iowa City, USA.

Bäch, T. & Kursawe, F. 1995. "Evolutionary algorithms for fuzzy logic: a brief review" in *Advances in fuzzy systems – applications & theory*, Vol. 4 (Fuzzy logic and soft computing ed. B. Bouchon-Meunier *et al.*) No. 1. World Scientific, London.

Bárdossy, A. & Duckstein, L. 1995. *Fuzzy rule-based modelling with applications to geophysical, biological and engineering systems*. CRC press Inc.

Bárdossy, A., Bogardi, I. & Duckstein, L. 1990. Fuzzy regression in hydrology, *Water Resources Research*, 26 (7) 1497-1508.

Bárdossy, A., Bronstert, A. & Merz, B. 1995. 1-, 2- and 3-dimensional modeling of water movement in the unsaturated soil matrix using a fuzzy approach, *Adv. Wat. Resour.* 18 (4) 237-251.

Beven, K.J & Binley, A. 1992. The future of distributed models: model calibration and uncertainty prediction, *Hydrol. Processes*, 6, 279–298.

Bishop, C.M. 1995. *Neural networks for pattern recognition*. Oxford University Press, Oxford, UK.

Boden, M.A. 1977. *Artificial intelligence and natural man*, Basic Books, New York.

Campolo, M., Soldati, A & Andreussi, P. 2003. Artificial neural network approach to flood forecasting in the River Arno, *Hydrol. Sc. J.* 48 (3) 381–398.

Canizares, R. 1999. *On the application of data assimilation in regional coastal models*. PhD Thesis, IHE Delft, Balkema, Rotterdam, The Netherlands.

Carpa, A., Nicosia, O.L.D., & Scicolone, B. 1994. Application of fuzzy sets to drought assessment, *Adv. in Wat. Resour. Tech. and Management*, (ed. Tsakiris & Santos), Balkema, 479-483.

Carsel, R.F., Parrish, R.S., Jones, R.L., Hansen, J.L. & Lamb, R.L. 1988. Characterising the uncertainty of pesticide leaching in agricultural soils, *J. of Contaminant Hydrology*, 2, 111-124.

Chaitin, G.J. 1987. *Information, randomness, and incompleteness: Papers on Algorithmic Information Theory*. World Scientific, Singapore.

Chiu, C.L. (ed.) 1978. *Application of Kalman filter to hydrology, hydraulics and water resources*, 1st ed., Univ. of Pittsburgh.

Chow, V.T., Maidment, D.R. & Mays, L.W. 1988. *Applied hydrology*. McGraw-Hill, New York.

Cunge, J.A. 1969. On the subject of flood propagation computation method (Muskingum method), *J. of Hydraulic Research*, 7 (2) 205-230.

Cunge, J.A. 2003. Of data and models, *J. of Hydroinformatics,* 5 (2) 75-98.

Cunge, J.A., Holly Jr, F.M. & Verwey, A. 1980. *Practical aspects of computational river hydraulics.* Pitman, London.

Cybenko, G. 1989. Approximations by superposition of a sigmoidal function, *Mathematics of Control, Signals and Systems,* 2, 303-314.

Daw, C.S., Finney, C.E.A. & Tracy, E.R. 2003. A review of symbolic analysis of experimental data, *Review of Scientific Instruments,* 74, 916-930.

Dawson, C.W. & Wilby, R. 1998. An artificial neural network approach to rainfall–runoff modelling. *Hydrol. Sc. J.* 43 (1) 47–66.

Dibike, Y.B., Solomatine, D.P. & Abbott, M.B. 1998. On the encapsulation of numerical-hydraulic models in artificial neural networks, *J. of Hydraulic Research,* 37 (2) 147-162.

Dibike, Y. & Solomatine, D.P. 1999. River flow forecasting using artificial neural networks, *European Geophysical Society,* XXIV general assembly, The Hague, The Netherlands.

Droste, C. 1998. *Uncertainty in parameter estimation for nonlinear dynamical systems.* German Geodetic Commission, Munich, Germany.

Dubois, D. and Prade, H. 1988. "An introduction to Possibilistic and Fuzzy Logics" in *Non-Standard Logics for Automated Reasoning* (ed. P. Smits, *et al.*), Academic Press Limited, London.

Fontane, D.G., Timothy, K.G. & Moncado, E. 1997. Planning reservoir operations with imprecise objectives, *J. of Wat. Resour. Planning and Management, ASCE,* 123 (3) 154-162.

Franchini, M. & Pacciani, M. 1991. Comparative analysis of several conceptual rainfall–runoff models, *J. of Hydrology,* 122, 161–219.

Franchini, M. 1996. Use of a genetic algorithm combined with a local search method for the automatic calibration of conceptual rainfall–runoff models, *Hydrol. Sc. J.* 41 (1) 21–39.

Franchini, M., Galeati, G. & Berra, S. 1998. Global optimisation techniques for the calibration of conceptual rainfall–runoff models, *Hydrol. Sc. J.* 43 (3) 443–458.

Franchini, M., Wendling, J., Obled, C. & Todini, E. 1996. Physical interpretation and sensitivity analysis of the TOPMODEL, *J. of Hydrology,* 175, 293–338.

Fraser, A.M. & Swiney, H.L. 1986. Independent coordinates for strange attractors from mutual information, *Physical Review,* 33 (A) 1134-1140.

Frison, T. 2000. Dynamics of the residuals in estuary water levels, *Physics and Chemistry of the Earth,* Part B, 25 (4) 359-364.

Frison, T., Abarbanel, H., Earle, M., Schultz, J. & Sheerer, W. 1999. Chaos and predictability in ocean water level measurements, *J. of Geophysical Review,* 104 (C4) 7935 - 7951.

Gelhar, L.W., Welty, C. & Rehfeldt, K.R. 1992. A critical review of data on field-scale dispersion in aquifers, *Water Resources Research,* 28 (7) 1955-1974.

Gerritsen, H., De Vries, J.W. & Phillippart, M.E. 1995 "The Dutch Continental Shelf Model" in *Coastal and Estuarine Studies: Quantitative Skill Assessment for Coastal Ocean Models,* (ed. D.R. Lynch & A. M. Davies), American Geophysical Union, 47, 425-467.

Goldberg, D.E. 1989. *Genetic algorithms in search, optimization and machine learning.* Addison-Wesley, Reading, Massachusetts.

Guinot, V. 1995. *Modélisation mécaniste du devenir des produits phytosanitaires dans l'environnement souterrain. Application à la protection des captages en aquifère*. PhD Thesis, University of Grenoble, France.

Guinot, V. 2001. The discontinuous profile method for simulating two-phase flow in pipes using the single component approximation, *Int. J. for Num. Meth. in Fluids*, 37 (3) 341-359.

Hall, J.W. & Davis, J.P. 1998. Process modelling and decision support for flood defence systems, *Proc. Hydroinformatics '98*, 24-26 August 1998, Copenhagen, Denmark (ed. V.M. Babovic & L.C. Larsen), Balkema, Rotterdam, 1, 285-292.

Hall, J.W. 1999. *Uncertainty management for coastal defence systems*. PhD thesis. Department of Civil Engineering, Bristol University, UK.

Hall, M.J. 2001. How well does your model fit the data?, *J. of Hydroinformatics*, 3 (1) 49-55.

Haykin, S. 1994. *Neural networks: A comprehensive foundation*. Macmillan, New York.

Hecht-Nielson, R. 1991. *Neurocomputing*. Addison-Wesley Pub. Co.

Henderson, F.M. 1966. *Open channel flow*. MacMillan, New York.

Herrera, F., Lozano, M. & Verdegay, J. L. 1995. "Generating fuzzy rules from examples using genetic algorithms" in *Advances in fuzzy systems – applications & theory*, Vol. 4 (Fuzzy logic and soft computing ed. B. Bouchon-Meunier *et al.*) No. 2. World Scientific, London.

Hornik, K., Stinchcombe, M. & White, H. 1989. Multilayer feedforward networks are universal function approximators, *Neural Networks*, 4, 251-257.

Imrie, C.E., Durucan, S. & Korre, A. 2000. River flow prediction using artificial neural networks: generalisation beyond the calibration range, *J. of Hydrology*, 233, 138–153.

Ishibuchi, H., Nozaki, K. & Tanaka, H. 1992. Distributed representation of fuzzy rules and its application to pattern classification, *Fuzzy Sets and Systems*, 52, 21-32.

Ishibuchi, H., Nozaki, K., Yamamoto, N. & Tanaka, H. 1995. Selecting fuzzy if-then rules for classification problems using genetic algorithms, *IEEE Transactions on Fuzzy Systems*, 3(3) 260-270.

Jaynes, E.T. 1982. On the rationale of maximum entropy methods, *Proc. IEEE*, 70 (9) 939-952.

Kalman, R.E. 1960. A new approach to linear filtering and prediction problems, *Trans. ASME, J. of Basic Engineering*, 82, 34-45.

Kantz, H. & Schreiber, T. 1997. *Nonlinear time series analysis*. Cambridge University Press.

Kasabov, N.K. 1996. *Foundations of neural networks, fuzzy systems, and knowledge engineering*. MIT Press.

Klir, G.J & Folger, T.A. 1988. *Fuzzy sets, uncertainty and information*. Prentice-Hall, London.

Klir, G.J. & Wierman, M.J. 1998. *Uncertainty-based information: elements of generalized information theory*. Springer-Verlag, Heidelberg, Germany.

Klir, G.J. 1990. A principle of uncertainty and information invariance, *Int. J. of General Systems*, 17 (2&3) 249-275.

Klir, G.J. 1994. "The many faces of uncertainty" in *Uncertainty modeling and analysis: theory and applications* (ed. B.M Ayyub & M.M. Gupta), 3-19, Elsevier Science.

Kohonen, T. 1990. The self-organizing map, *Proc. of the IEEE*, 78, 1464-1497.

Kolmogorov, A.N. 1965. Three approaches to the qualitative definition to information, *Problems of Information Transmission*, 1 (1) 1-7.

Kosko, B. 1992. *Neural networks and fuzzy systems: a dynamical approach to machine intelligence.* Englewood Cliffs, NJ, Prentice-Hall.

Kosko, B. 1993. *Fuzzy thinking, the new science of fuzzy logic.* Flamingo, HarperCollins Publ., London.

Laviolette, M. & Seaman, J.W. 1994. The efficacy of fuzzy representations of uncertainty, *IEEE Transactions on Fuzzy Systems,* 2 (1) 4-15.

Lee, Y.H. & Singh, V.P. 1998. Application of the Kalman filter to the Nash model, *Hydrol. Processes,* 12 (5) 755-767.

Lekkas, D.F., Imrie, C.E. & Lees, M.J. 2001. Improved non-linear transfer function and neural network methods of flow routing for real-time forecasting, *J. of Hydroinformatics,* 3 (3) 153-164.

Li, H.X. & Yen, V.C. 1995. *Fuzzy Sets and Fuzzy Decision-Making.* CRC Press Inc.

Linsley, R.K., Kohler, M.A. & Paulhus, J.L.H. 1988. *Hydrology for Engineers.* McGraw-Hill, Singapore.

Maier, H.R. & Dandy, G.C. 2000. Neural networks for the prediction and forecasting of water resources variables: a review of modelling issues and applications, *Environmental Modelling & Software*, 15, 101–124.

Mamdani, E.H. & Assilian, S. 1975. An experiment in linguistic synthesis with a fuzzy logic controller, *Int. J. of Man-Machine Studies*, 7 (1) 1-13.

McCarthy, G.T. 1938. The unit hydrograph and flood routing, *Proc. conf. of the North Atlantic Division of the US Corps of Engineers*, 24 June 1938.

McCulloch, W.S. & Pitts, W. 1943. A logical calculus of the ideas immanent in nervous activity, *Bulletin of Mathematical Biophysics*, 5, 115-137.

Minns, A.W & Hall, M.J. 1995. Artificial neural networks as rainfall-runoff models, *Hydrol. Sc. J.* 41 (3) 399-417.

Minns, A.W. 1998. *Artificial neural networks as subsymbolic process descriptors.* PhD Thesis, IHE Delft, Balkema, Rotterdam.

Mohamed Nanseer, N.K. 2003. *Determination of confidence limits for model estimation using resampling techniques.* M.Sc. Thesis HH 448, IHE Delft, The Netherlands.

Moody, T. & Darken, C. 1989. Fast learning in networks of locally tuned processing units, *Neural Computation*, 1, 281-294.

Mroczkowski, M., Raper, G.P. & Kuczera, G. 1997. The quest for more powerful validation of conceptual catchment models, *Wat. Resour. Res.*, 33 (10) 2325-2335.

Nash, J.E. & Sutcliffe, J.V. 1970. River flow forecasting through conceptual models, Part 1, A discussion of principles, *J. of Hydrology,* 10, 282–290.

Nash, J.E. 1959. Systematic determination of unit hydrograph parameters, *J. of Geophysical Research*, 64, 111-115.

Negnevitsky, M. 2002. *Artificial intelligence: a guide to intelligent systems*. Pearson Education Ltd.

O'Connell, P.E. & Clarke, R.T. 1981. Adaptative hydrological forecasting - a review, *Hydrol. Sc. Bull.* 26 (2) 179-205.

Pesti, G., Shrestha, B.P., Duckstein, L. & Bogárdi, I. 1996. A fuzzy rule-based approach to drought assessment, *Wat. Resour. Res.*, 32 (6) 1741-1747.

Price, R.K. 1973. Flood routing methods for British rivers, *Inst. Civ. Engineers,* London, 55, 913-930.

Price, R.K. 1976. Equivalent river models, *proc. Int. Symp. on Unsteady Flow in Open Channels*, BHRA, Paper K4, Newcastle.

Price, R.K. 1985. "Flood routing" in *Developments in Hydraulic Engineering* (ed. P. Novak), Elsevier Applied Science Publishers, London, 129-174.

Rajurkar, M.P., Kothyari, U.C. & Chaube, U.C. 2002. Artificial neural networks for daily rainfall-runoff modeling, *Hydrol. Sc. J.* 47 (6) 865-877.

Refsgaard, J.C. 1997. Validation and intercomparison of different updating procedures for real-time forecasting, *Nordic Hydrology,* 28, 65-84.

RIZA 1993. *The River Rhine*. Directorate-General for Public Works and Water Management, The Netherlands. RIZA Institute for Inland Water Management and Waste Water Treatment.

Romanowicz, R & Beven, K. 1998. Dynamic real-time prediction of flood inundation probabilities, *Hydrol. Sc. J.* 43 (2) 181–196.

Rosenblatt, F. 1958. The perceptron: a probabilistic model for information storage and organization in the brain, *Psychological Review*, 65, 386-408.

Rumelhart, D.E., Hinton, G.E. & Williams, R.J. 1986. "Learning internal representations by error propagation" in *Parallel distributed processing: explorations in the microstructure of cognition* (ed. D.E. Rumelhart & J.L. McClelland), 1, 318-362, MIT Press, Cambridge, MA.

Schulz, K. & Huwe, B. 1997. Water flow modeling in the unsaturated zone with imprecise parameters using a fuzzy approach, *J. of Hydrology,* 201, 211-229.

Schulz, K. & Huwe, B. 1999. Uncertainty and sensitivity analysis of water transport modeling in a layered soil profile using fuzzy set theory, *J. of Hydroinformatics,* 1 (2) 127-138, IWA Publishing.

See, L. & Openshaw, S. 1999. Applying soft computing approaches to river level forecasting, *Hydrol. Sc. J.* 44 (5) 763–778.

Serban, P. & Askew, A.J. 1991. Hydrological forecasting and updating procedures, *IAHS Pub.,* 201, 357–369.

Shafer, G. 1976. *A mathematical theory of evidence*. Princeton University Press, Princeton, New Jersey.

Shamseldin, A.Y. & O'Connor, K.M. 2001. A non-linear neural network technique for updating of river flow forecasts, *Hydrology and Earth System Sc.,* 5 (4) 557–597.

Shamseldin, A.Y., O'Connor, K.M. & Liang, G.C. 1999. Methods for combining the outputs of different rainfall–runoff models, *J. of Hydrology,* 197, 203–229.

Shannon, C.E. 1948. A mathematical theory of communication, *Bell System Technical J.* 27, 379-423 and 623-656.

Shynk, J.J. 1990. Performance surfaces of a single-layer perceptron, *IEEE Transactions on Neural Networks,* 1, 268-274.

Silva, W., Klijn, F. & Dijkman, J. 2001 (ed.). *Room for the Rhine branches in The Netherlands: what research has taught us.* Directorate-General for Public Works and Water Management, The Netherlands.

Singh, V.P. 1998. *Entropy-based parameter estimation in hydrology.* Kluwer Academic Publishers, Dordrecht, The Netherlands.

Solomatine, D.P. & Dulal, K.N. 2003. Model trees as an alternative to neural networks in rainfall-runoff modeling, *Hydrol. Sc. J.* 48 (3) 399–411.

Solomatine, D.P. 1999. Two strategies of adaptive cluster covering with descent and their comparison to other algorithms, *J. of Global Optimization,* 14 (1) 55–78.

Solomatine, D.P., Velickov, S. & Wust, J.C. 2001. Predicting water levels and currents in the North Sea using chaos theory and neural networks, *Proc. XXIX IAHR Congress,* 16-21 September 2001, Beijing, China.

Sriwongsitanon, N., Ball, J.E. & Cordery, I. 1998. An investigation of the relationship between the flood wave speed and parameters in runoff-routing models, *Hydrol. Sc. J.* 43 (2) 197-213.

Stein, M. 1987. Large sample properties of simulations using Latin Hypercube Sampling, *Technometrics,* 29 (2) 143-151.

Sugeno, M. 1985. *Industrial applications of fuzzy control.* North Holland, Amsterdam.

Tingsanchali, T & Manusthiparom, C. 2001. A neural network model for flood forecasting in tidal rivers, *Proc. XXIX IAHR congress,* 16-21 September 2001, Beijing, C, 85-92.

Todini, E. 1996. The ARNO rainfall–runoff model, *J. of Hydrology,* 175, 339–382.

Uhlenbrook, S., Seibert, J., Leibundgut, C. & Rodhe, A. 1999. Prediction uncertainty of conceptual rainfall–runoff models caused by problems in identifying model parameters and structure, *Hydrol. Sc. J.* 44 (5) 779–797.

Varoonchoticul, P., Hall, M.J. & Minns, A.W. 2002. Extrapolation management for artificial neural network models of rainfall-runoff relationships, *Proc. Hydroinformatics 2002,* 1-5 July 2002, Cardiff, UK, IWA Publishing, 2, 673-678.

Velickov, S. 2002. Chaos and predictability in the North Sea water levels along the Dutch coast, *Physics and Chemistry of the Earth, Part B: Hydrology, Oceans and Atmosphere.*

Verlaan, M. 1998. *Efficient Kalman filtering algorithms for hydrodynamic models.* PhD Thesis, Delft University of Technology, The Netherlands.

Welstead, S.T. 1994. *Neural network and fuzzy logic applications in C/C++.* Wiley Professional Computing, New York.

WL | Delft Hydraulics. 2001. *Discharge-boundary-conditions of the Zeedelta model.* Report number Q2892, April 2001.

Wong, T.H.F. & Laurenson, E.M. 1983. Wave speed-discharge relations in natural rivers, *Wat. Resour. Research*, 19, 701-706.

Yang, X. & Michel, C. 2000. Flood forecasting with a watershed model: a new method of parameter updating, *Hydrol. Sc. J.* 45 (4) 537-546.

Young, P. 2001. "Data-based mechanistic modelling and validation of rainfall-flow processes" in *Model validation: perspectives in hydrological science* (ed. M.G. Anderson & P.D. Bates), 118-161, J. Wiley and Sons, Chichester, UK.

Zadeh, L.A. 1965. Fuzzy sets, *Information and control,* 8, 338-353.

Zadeh, L.A. 1975. The concept of a linguistic variable and its application to approximate reasoning, part I, *Information Sciences*, 8, 199-249.

Zadeh, L.A. 1983. The role of Fuzzy Logic in the management of uncertainty in expert systems, *Fuzzy sets and systems*, 11, 199-227.

Zhao, R.J., Zhuang, L.R., Fang, X., Liu, R. & Zhang, Q.S. 1980. "The Xinanjiang model" in *Hydrological Forecasting*, (Proc. Oxford Symp., April 1980), 351–356. IAHS Publ. no. 129.

Zijderveld, A. 2003. *Neural network design strategies and modelling in Hydroinformatics.* PhD Thesis, Delft University of Technology, The Netherlands.

Curriculum Vitae

Abebe Andualem Jemberie was born on 22 December 1971 in Merawi, Ethiopia. In 1992 he received a B.Sc. degree in Hydraulic Engineering with distinction from Arbaminch Water Technology Institute (AWTI), Arbaminch, Ethiopia. From October 1992 to September 1996 he worked as a graduate assistant, an assistant lecturer and then a lecturer in the Hydraulic Engineering department of AWTI. In October 1996 he joined the postgraduate course in Hydraulic Engineering at the International Institute for Infrastructural, Hydraulic and Environmental Engineering (IHE), Delft, The Netherlands and in April 1998, he obtained his M.Sc. degree in Hydroinformatics with distinction. Since April 1998, he continued to work at IHE as a teaching and research assistant and a PhD student.

T - #0285 - 101024 - C0 - 246/174/11 - PB - 9789058096951 - Gloss Lamination